高等院校土建学科双语教材（中英文对照）
◆土木工程专业◆
BASICS

建筑立面开洞
FAÇADE APERTURES

[德] 罗兰·克里普纳（Roland Krippner） 编著
弗洛里安·穆索（Florian Musso）
杨 璐 译

中国建筑工业出版社

著作权合同登记图字：01-2009-7704号

图书在版编目（CIP）数据

建筑立面开洞/（德）克里普纳，（德）穆索编著；杨璐译 .—北京：中国建筑工业出版社，2013.5

高等院校土建学科双语教材（中英文对照）◆土木工程专业◆

ISBN 978-7-112-15361-9

Ⅰ.①建… Ⅱ.①克…②穆…③杨… Ⅲ.①窗-建筑设计-高等学校-教材-汉、英②窗-建筑工程-工程施工-高等学校-教材-汉、英 Ⅳ.①TU228

中国版本图书馆 CIP 数据核字（2013）第 085399 号

Basics: Façade Apertures/ Roland Krippner, Florian Musso
Copyright © 2008 Birkhäuser Verlag AG, P. O. Box 133, CH-4010 Basel, Switzerland
Chinese Translation Copyright © 2013 China Architecture & Building Press
All rights reserved.
本书经Birkhäuser Verlag AG出版社授权我社翻译出版

责任编辑：孙书妍　　责任设计：陈旭　　责任校对：张 颖 赵 颖

高等院校土建学科双语教材（中英文对照）
◆土木工程专业◆
建筑立面开洞
[德]　罗兰·克里普纳（Roland Krippner）
　　　弗洛里安·穆索（Florian Musso）　编著
　　　杨璐　　　　　　　　　　　　　　译
*
中国建筑工业出版社出版、发行（北京西郊百万庄）
各地新华书店、建筑书店经销
北京嘉泰利德公司制版
北京云浩印刷有限责任公司印刷
*
开本：880×1230 毫米　1/32　印张：4⅞　字数：140 千字
2013 年 7 月第一版　2013 年 7 月第一次印刷
定价：18.00 元
ISBN 978-7-112-15361-9
　　　（23402）

版权所有　翻印必究
如有印装质量问题，可寄本社退换
（邮政编码 100037）

中文部分目录

\\ 序　7

\\ 前言　87

\\ 开洞的功能　91
　　\\ 保护功能　91
　　\\ 控制功能　93
　　\\ 边缘构件　101
　　\\ 控制设备　104

\\ 洞口的组成　113
　　\\ 构件　113
　　\\ 基本构造　116

\\ 窗户的组成　121
　　\\ 开启方式　121
　　\\ 构造类型　125
　　\\ 窗框　126
　　\\ 材料——系统窗户　128
　　　　\\ 木窗　128
　　　　\\ 金属窗　129
　　　　\\ 塑料窗（PVC：聚氯乙烯，
　　　　　　GRP：玻璃增强塑料、玻璃钢）　132
　　\\ 玻璃系统　132
　　　　\\ 单层玻璃　133
　　　　\\ 热处理玻璃　136
　　　　\\ 多层玻璃　136
　　　　\\ 功能玻璃　137
　　　　\\ 特殊玻璃　137
　　\\ 紧固系统　142
　　\\ 控制设备　142

\\ 结构构件的整体安装　144
　　\\ 窗户底部　144
　　\\ 密封做法　145
　　\\ 接缝　147

\\ 结语　152

\\ 附录　153
　　\\ 参考文献　153
　　\\ 图片鸣谢　154
　　\\ 作者简介　155

TABLE OF CONTENTS

\\Foreword _9

\\Introduction _11

\\Aperture functionality _15
    \\Protective functions _15
    \\Control functions _18
    \\Edges _26
    \\Control devices _29

\\Aperture components _39
    \\Elements _39
    \\Basic constructions _43

\\Window components _48
    \\Opening types _48
    \\Construction types _52
    \\Frames _55
    \\Materials – system windows _56
        \\Timber windows _56
        \\Metal windows _57
        \\Plastic windows (PVC, GRP) _60
    \\Glazing systems _60
        \\Single-layer glass _61
        \\Thermally treated glass _63
        \\Multi-layered glass _64
        \\Additional-leaf glass _65
        \\Special glass _65
    \\Fastening systems _70
    \\Fittings _71

\\Fitting structural elements together _73
    \\Bottom of the window _73
    \\Seals _74
    \\Joints _77

\\In conclusion _81

\\Appendix _83
  \\Literature _83
  \\Picture credits _84
  \\The authors _85

序

建筑的外表面需要满足建筑所需的一些功能。首先，建筑外表面需要保护建筑内部避免受到外力和外部作用的影响——包括天气、温差以及防止从外部透视。通常情况下，建筑外表面并不需要对建筑进行完全密封，在保护建筑内部的同时还需要向外部空间相互开放，以进行空气交换和采光。建筑开洞从功能方面和（或）视觉方面将建筑内部和外部进行了相互联系。

"高等院校土建学科双语教材"系列丛书中《建筑立面开洞》一册的主要内容是针对建筑立面开洞的建造施工设计。本册首先详细地介绍了窗户所应该具有的功能，然后讨论了窗户建造过程中结构方面的相关细节。主要内容包括：窗户的不同组成构件、结构类型和材料，以及各自特点的介绍；阐述窗户与周围墙体的连接，特别是如何实现窗户和周围墙体开洞的过渡；多层窗户的不同功能介绍。房间的采光来自于墙体的竖向开洞的外侧。本册的重点集中在窗户的介绍上，而对于门和固定窗户，不论开洞的尺寸大小，其设计原理均与窗户相同。

从结构、功能和设计角度而言，立面开洞的四周部位与开洞本身的重要性相同。本册还对开洞周围采光、通风以及隔热设施和控制设备进行了介绍。

对于建筑师而言，应该熟知用于设计窗户、玻璃门的结构和构件，才能将设计出的建筑立面开洞具有独特的个性，同时使建筑内、外部和谐一致。

编辑：贝尔特·比勒费尔德（Bert Bielefeld）

FOREWORD

The outer skin of a building fulfils a number of functions. It protects the interior from external influences – weather, temperature differences, or from being overlooked. It is not usually desirable to be sealed off hermetically, so that as well as offering protection, the facade must open up to the outside world, interact with it and admit air and light. Openings link the inside and the outside functionally and/or visually, and relate to the two to each other.

The present volume in the Basics series for students deals with designing facade openings in the general field of construction. The many demands made on a window are first explained clearly, and the structural detail needed to construct them is then discussed. The various components, structural types and materials are presented with their specific characteristics and tied into the context of the surrounding wall areas, with close attention paid to understanding how the transition from opening to wall should be achieved, and the way in which the different functions of a window are operated in strata. These observations relate to rooms lit from one side with apertures placed vertically in the wall. Here the focus is on windows; regardless of the size of the opening, similar demands are made on doors and fixed glazing, and the same principles applied.

The peripheral areas of a facade are important structurally, functionally and in terms of design, as well as the actual openings. Devices for controlling the light admitted, ventilation and heat insulation around the opening are also discussed.

Familiarity with the structures and components that can be used for designing windows and glazed doors is an important basis for design work by architects who wish to give facades an individual character and create a composition that is harmonious both inside and out.

Bert Bielefeld
Editor

INTRODUCTION

An opening or aperture is generally defined as an "open place, hole, gap". In a building, openings are defined as empty spaces left in the wall. If creating a protected space is the building's first priority, then the openings in the facades are the next, essential measure if the space is to be used.

Furthermore, openings are an essential architectural design element. Their dimensions and proportions, their position in relation to the water-bearing layer (waterproof skin) or to the surface concluding a space, their arrangement and relation to each other make a crucial difference to the design of buildings. The elements used to close the aperture are either movable, depending on their function (windows, doors; both are basic functional building blocks), or fixed (glazing).

Windows play a central part in planning openings. Their form (components, formats), arrangement and distribution over the facade are the building's visiting card. The division of the windows and the internal structuring of the glazed area is another important feature within the overall effect. Stylistic developments and craft skills are also clearly discernible in window forms.

Control devices are additional systems for apertures, beyond the doors and windows themselves. They make it possible to control permeability for light, air and heat precisely. The degree of comfort in the interior can be adapted to suit the level of demand and users' needs, according to the weather conditions.

A wide range of structural elements and systems are available. Familiarity with basic functional principles and general structural conditions makes it possible to develop effective strategies for planning openings in relation to a particular climatic situation. Here the design aspects interact closely with functional requirements and structural qualities.

Openings perform the same protective functions (against cold, damp, noise, fire and intruders) as the facade, but are thermal weak points in the building envelope, and require a new strategic approach given tightened energy requirements. Openings also create possibilities of access and define areas in the wall concluding the space that provide the room with light and ventilation, and makes it possible to look out from the inside.

Fig.1: Opening in natural masonry

Fig.2: Opening in solid timber wall (log construction)

Fig.3: Round window (in a 1950s pavilion)

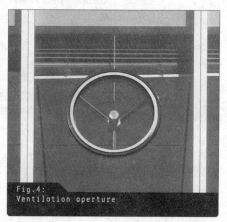
Fig.4: Ventilation aperture

Fig.5: Story-height openings

Fig.6: Honeycomb openings

It therefore has to be possible to alter the permeability of these openings, so that the effects of heat, light and air can be regulated by special closing elements, appropriate to the location (changing climatic conditions) and the need for comfort (constant interior climate). The quality profile depends on position in relation to the sun, on considerations relating to the interior space, and on utilization.

Doors, which are older elements than windows in terms of architectural history, provide access to the building. They also include gates, although these are large and usually intended to be driven through. In most people's minds, openings in the facade more or less equate to windows. These are the elements that allow light to pass through transparent or translucent material even when they are closed, and allow air to flow in and out when the individual windows are open.

The simplest form of an element that closes apertures is fixed glazing. Lighting and ventilation can also be achieved separately from each other (e.g. a combination of window glazing and separately operated ventilation flaps).

Openings have a public aspect, facing outwards, which meets some of the demands made by prestige. Their number and size suggest the social status of owner and occupants. The inward-facing side has a character of privacy and intimacy. The arrangement of doors and windows influences the effect made by the interior as well. Rooms that are well provided with windows can seem more generous and airy than the real dimensions would convey.

APERTURE FUNCTIONALITY

PROTECTIVE FUNCTIONS

The basic needs that windows meet have remained largely the same for centuries, but there has been a considerable change in requirements for thermal and sound insulation and fireproofing, and for airtightness where the structural elements meet the wall. The demands made on a window's properties or efficiency depend on the location of the building – the topography, the direction in which it faces, and the height of the building, and are strongly affected by the prevailing wind.

Heat and sound insulation

The heat and sound insulation a window provides depend essentially on the material chosen for the frame and its structure or thickness, the nature and thickness of the glazing, and the way the component is fitted.

A window has to meet minimum heat insulation requirements and thus protect the interior against cold. As elements, frame and glass must be able to buffer temperature peaks between the interior and the exterior despite the relative thinness of the structural element. To avoid heat bridges the window should be positioned within the insulating layer of the wall as far as possible; and to avoid condensation and mold the interior surfaces should be at high temperatures. › Chapter Window components, Glazing systems

Spaces adjacent to the exterior wall should be insulated from the outside world in terms of sound. In most cases the interior should be protected from exterior noise (e.g. aircraft, road traffic, etc.), but the reverse is also true.

The required heat and sound insulation can be achieved by using special glass (heat and sound insulation glazing), or by the type of window

---

\\ Note:
In exterior walls and glazing in particular, the interior surface temperature affects the heat requirements and the sense of comfort. The temperature that users perceive is an average of air and surface temperature. Choice of material, wall structure and/or improved windows should be used to keep interior surface temperatures within the perceived temperature range.

\\ Note:
Accumulated condensation
If the water vapor in the (room) air cools to the extent that it becomes liquid, condensation forms. Surface temperatures should thus be kept as high as possible, especially for glazing and attachment points.

construction (winter, double-glazed, box-type windows). Improved heat insulation raises the surface temperatures for the glass pane on the interior side. This reduces the physiologically unpleasant presence of cold air in the window area. Better glass increases the demands made on the structural window components in terms of heating technology. Laminated, insulated timber frames or thermally separated metal elements improve the overall qualities of windows considerably.

Attention must be paid to protection against summer heat or overheating, as well as to winter heat insulation. This depends primarily on the size of the opening, as well as on the way the building faces.

Damp protection and airtightness

Windows should not let in rain (and splash water) or dampness, as this would damage the fabric of the building.

To avoid drafts and ensure controlled ventilation the windows should have airtight fittings and fastenings. This applies above all to the joint with the building, but also to the one between the fixed frame and the opening section. Improved insulation standards diminish the impact of ventilation heat losses on the building's energy requirements.

Moisture from the interior affects the windows and their joints, along with the effects of the weather in general. In relation to water vapor diffusion, the principle applies that the functional layers should always be thicker inside than they are outside. This ensure perfect moisture transport through the building structure from the interior.

Screens, glare protection

Large apertures need sight screening to ensure the privacy of the spaces behind them.

Glare is caused by marked light density contrasts, which are particularly disturbing in workplaces with monitors. Antiglare systems control

\\ Important:
It is essential for ergonomic working that windows and skylights be constructed or provided with antiglare or sun protection devices so that rooms can be screened from direct exposure to the sun. Regulations or recommendations can be found in EU Directive 90/270/EEC, ISO 9241, and in a whole range of national codes.

Fig.7:
Semitransparent antiglare system

Fig.8:
Protection against falling

inward radiation, thus reducing the brightness differential between field of vision and computer screen.

Antiglare devices are place on the interior side. They should not completely block out daylight or make visual contact impossible. Systems that can be raised from the lower edge of the opening and positioned as wished without covering the whole surface of the glass are particularly suitable. > Fig. 7 Sight-screening can be placed outside, inside an additional-leaf window, or inside the room.

Fire protection

Fire must not be allowed to break out in buildings, but if it does, the fire must not be allowed to spread. Fires can spread via apertures from room to room, and from story to story. Window frames and glazing are subject to fire protection requirements to limit fire spread. These have to be met by the frame materials and the glazing selected.

Impact resistance, fall protection

In the case of story-high glazing in particular (shop windows, panoramic panes, etc.), window design can be affected by possible impact load, making special glass of additional retention structures necessary.

The window breast height required by building regulations (depending on the maximum acceptable drop height) must be met to prevent falls from all opening windows and for story-high, large-format fixed glazing and or fully glazed doors. > Chapter Aperture components, Elements This can be achieved with a window breast, a fixed glazing element or a rail. > Fig. 8

Burglar resistance

Easily accessible areas of the exterior wall (e.g. the ground floor or windows linked by rendered balconies) usually need special measures for protection against burglary. Windows and glazed doors are at risk from being opened by simple prying tools, as normal window fittings are not burglar-proof. Protection against break-ins can be improved by installing a matching overall structure to resist burglars (frame complete with opening element, fittings and glazing), or by using safety or window grilles or opening devices.

CONTROL FUNCTIONS

When planning openings – their size, arrangement, etc. – a variety of requirements, some contradictory, must be met and combined coherently.

The aperture size affects not just the degree of view through the opening and contact with the outside world, but above all the lighting possibilities and the direct use of solar energy.

Aperture area

Orientation towards a point of the compass is crucially important, along with absolute size, the "aperture area" (area of opening, proportion of frame).

Increasing the aperture size means:

_ More daylight admitted
_ More radiation admitted
_ Overheating problems in summer
_ Reduced heat insulation
_ Increased need for cleaning

Arrangement within the wall surface and the aperture's geometrical shape are always linked with the room behind. Both affect the amount of light admitted, ventilation and the users' visual connections with the outside world.

\\ Note:
Most countries regulate the arrangement of openings in walls that face adjacent buildings, stipulating separation distances, etc. As a rule the minimum distance apart is ≥ 5 m.

\\ Important:
The following dimensions can be used as a basis for planning sightlines in living rooms (height above floor):
_ Top edge of window area ≥ 2.2 m
_ Bottom edge of window area ≥ 0.95 m
_ Width of window area ≥ 55% of room width.

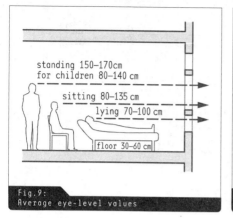

Fig. 9:
Average eye-level values

Fig. 10:
Bar-free area at eye level

The position of an opening relates to use both horizontally and vertically. Variants in furniture can change the horizontal relationship with openings.

On the vertical, the exterior wall is divided into the following areas:

- Transom light (top part of the window above the transom)
- Window light (field of vision from the interior)
- Window breast

Sightlines

Contact with the outside world is achieved by using transparent, distortion-free and ideally color-neutral building materials such as glass or plastic. Sightlines can be opened up through the size and arrangement of the windows – e.g. by using strip windows, panoramic panes, floor-to-ceiling glazing – and also strongly defined or restricted – e.g. by using smaller openings or a precisely placed arrangement. › Fig. 9

Eye level

People's eye level when sitting or standing should be considered when placing apertures. Human proportions such as body size, field of vision, and the particular activity (sitting, standing, lying), are key factors here. The following average eye-level or sightline values can be applied for various positions:

- approx. 150–170 cm when standing
- approx. 80–135 cm when sitting (at work)
- approx. 70–100 cm when lying down

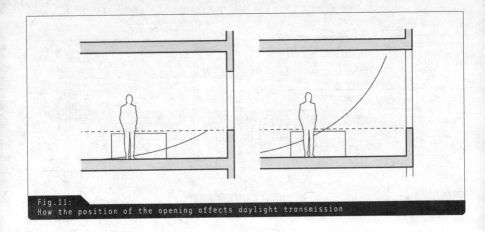

Fig.11:
How the position of the opening affects daylight transmission

The opening area should be subdivided according to the users' activities and positions, and the visual links with the outside space on the sightline should not be impaired by placing horizontal or vertical building components directly at eye level. > Fig. 10

Visual contact with the outside world is often associated with the desire for fresh air, i.e. with being able to stand by an open window. Opening and closing mechanisms should therefore be easily accessible.

Lighting

Natural lighting is important both physiologically and in terms of energy consumption. Daily and seasonal changes in the sun's position can be seen and experienced inside through the play of light and shade and the changing color of the light. Adequate natural light is important for well-being and good for productivity at work.

Building regulations generally require one-eighth of the floor space (usable space) as the minimum proportion for openings in living rooms. This minimum value is correlated with the depth of the room. If light

\\ Important:
The daylight factor (D) defines the proportion of incident light and gives the ratio of illumination strengths inside and outside (diffused light only) as a percentage under normal conditions.

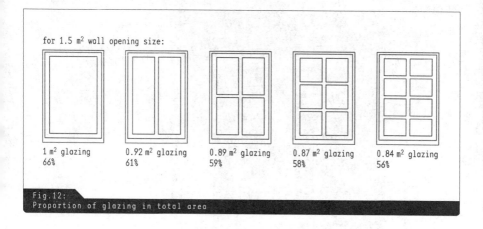

Fig.12:
Proportion of glazing in total area

enters from one side only, the amount of available light will diminish with the depth of the room.

External influences

External influences include:

_ Orientation and direction faced
_ Daily and seasonal variations
_ Lighting strength and light color
_ Shading from the immediate vicinity (vegetation, other buildings)

The position of the top edge of the aperture is important when lighting a room. The higher the window, the more light will penetrate to the back of the room. Given the same aperture area, the daylight factor will increase with the height of the top edge of the aperture above floor level. On the other hand, open areas in the window breast area (below the 0.85 m reference level) will make only a small contribution to improving incident daylight. > Fig. 11

The position of the opening in the outside wall is more important than absolute size for daylight utilization. Generally the amount of daylight penetrating through openings placed vertically in an outside wall is lower by a factor of approximately 5 than light from a horizontal opening of the same size in a flat roof. The degree of brightness actually available in a room depends on the degree of reflection from the interior surfaces, and this in turn is greatly influenced by dominant colors.

Workstations should be placed close to windows. Attention should be paid to the direction in which the light falls (to avoid people working

Fig. 13:
Art Nouveau window

in their own light) and the main direction in which the user looks (usually parallel to the opening).

When lighting a room, light can be lost through the positioning of structural elements, and through dirt. Up to 40% of the window area is taken up by frames, uprights, transoms, and bars. > Figs 12 and 13

Lateral structural elements, such as protecting wall sections, and elements that protrude above the opening (canopies, roof overhangs, balconies) restrict incident light and the way it is distributed in the room. The surfaces of such opaque structural elements should thus be as reflective as possible, i.e. they should be painted in light colors to throw the light back. Natural light can be exploited better by placing light-directing systems in front of openings and using highly reflective materials as ceiling cladding, to take light into the depths of the space.

Using solar energy

Sunlight is important for rooms in which leisure time is spent. Sunbeams coming in through glazed apertures can also be used to complement the building's heating system through the "greenhouse effect".

If combined with the spaces immediately behind them, openings are simple collection and storage systems. Four important parameters determine the proportion of solar energy that can be used directly: climatic and local factors, orientation in a particular direction, the aperture's angle of inclination, and freedom from shade.

In Central Europe, solar radiation is available but does not clearly coincide with heating needs on either a daily or a seasonal level. In a house with good heat insulation, up to one-third of the heat required can be supplied by solar energy if direct sunlight is exploited via south-facing surfaces in the heating period. If openings face east or west (receiving particularly intensive sunlight in summer), care should be taken to ensure that the aperture proportion is not above about 40%, in order to maintain thermal equilibrium.

The amount of radiation admitted is in direct proportion to the aperture area, so large glazed areas can cause overheating.

Exposure to sunlight, aperture size, heat needed and interior thermal storage materials should be in balance. Even in the heating period, openings taking up over 50% of the available area are not required. Provided that sunshading precautions are taken (e.g. projecting structural components), it is possible to achieve comfortable interior temperatures even without cooling systems.

Exploitable solar gain is additionally restricted by structural and user-related influences. Thus it makes sense to choose narrow frame sections and to use filigree subdivisions. Net or other curtains also reduce solar gain. Internal screening and antiglare systems further reduce the amount of solar radiation stored.

Actually exploitable solar radiation can be reduced by almost half the theoretical starting value as a result of building features, and can decrease further to one-third as a result of user-related influences.

Ventilation

After lighting, ventilation is most important aperture function. Ventilation is a basic condition for user wellbeing and health, and for protecting

---

\\ Important:
In offices with a high proportion of glazing, overheating in combination with internal thermal loads (artificial light, office equipment, people) means that a lot of cooling is needed.

\\ Note:
Solar gain refers to solar energy entering the building via windows and other transparent/translucent structural components. It helps to raise the temperature of the building and the air in the rooms, and thus reduce heating requirements. But only a proportion of the solar gain can be exploited, while the rest is dissipated in the surrounding area.

the building stock. Ventilation means exchanging the air in the room for fresh air. This traditionally happens via windows in the exterior wall area (desirable air change), or via joints between opening and window element or antiglare devices or opening leaves (undesirable air change).

Such air change is caused by interior and exterior pressure differences, and is called free or natural ventilation (also shock ventilation). Exchange of the internal air helps to supply the room with fresh air, remove water vapor, odors and exhaled CO_2, and contributes considerably to comfort levels.

Window ventilation

The window's area and the way it opens determine the degree of air change. Additional open windows (e.g. in an opposite wall), how far interior doors are open and the position of devices like roller or slatted blinds increase ventilation. As users behave in different ways, air change rates of 0.5 to 1/h are recommended. Air speed affects comfort as well as air change. The upper limit indoors is 0.2 m/s.

Windows that can be opened individually have the advantage that the fresh air admitted can be regulated directly by the user. Open windows should not knock against other structural elements (e.g. sunshading, antiglare or screening systems, wall, columns) or restrict the use of the interior. If windows are arranged on one side only, rooms are considered as suitable for natural ventilation if the room depth does not exceed the dimension of the clear opening height by a factor of more than 2.5.

Room ventilation is not necessarily linked to windows. In triple thermopane glazing in particular, or large-format panes, the weight of the fittings can make it preferable to separate a well-insulated, light ventilation flap from a fixed glazing element. > Fig. 14

\\ Note:
The air change rate n in the unit (1/h) indicates how often the room or building volume is changed within an hour.
Example: n = 10/h:
10 times the room or building volume is changed in one hour.

\\ Important:
Air change (AC)

Window positions	AC per hour
Window and door closed	0.1 to 0.3
Window tilted, blinds closed	0.3 to 1.5
Window tilted	0.8 to 4
Window half open	5 to 10
Window fully open	9 to 15
Opposite windows	
Intermediate doors fully open	up to 40

Fig.14:
Glazing element with opaque ventilation flap

Window ventilation raises a number of problems caused by user behavior. Regular ventilation is often neglected, and manual ventilation does not produce sufficient air change. As the flow of air cannot be controlled, this causes increased ventilation heat losses in winter and the need for additional cooling in summer.

Ventilation heat losses become more important for buildings' energy budgets as exterior wall construction improves in terms of heat technology. An inappropriate approach to ventilation can be avoided by a controlled air supply, but can lead to the loss of individual regulation. Mechanical inward and outward ventilation systems are installed in passive buildings in particular.

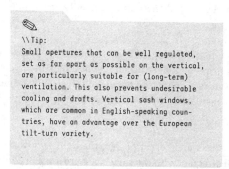

\\Tip:
Small apertures that can be well regulated, set as far apart as possible on the vertical, are particularly suitable for (long-term) ventilation. This also prevents undesirable cooling and drafts. Vertical sash windows, which are common in English-speaking countries, have an advantage over the European tilt-turn variety.

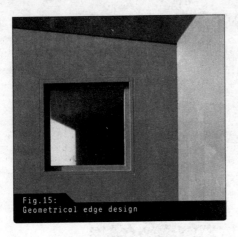

Fig.15:
Geometrical edge design

Air change through structural joints

In many (primarily older) buildings there is additional air change through structural joints. Joint ventilation cannot be controlled and does not distribute or remove air evenly. Joints can also cause damage to structural components by accumulated condensation. But this aspect is becoming less relevant as seals improve and interfaces can be made completely airtight in new buildings.

EDGES

Openings are bordered by structural elements that perform different functions according to position. A structure based on right angles, resulting in the first place from the building materials used and their modular dimensions, is not essential. Adaptations can affect daylight input, the view outside, and the direct use of solar energy.

In construction with thick walls in particular, the aperture is also the space between the outer, weatherproof skin and the internal surface wall, which can be shaped three-dimensionally according to the position of the window.

Energy-saving building and more modern approaches have also meant that walls are no longer becoming less thick in "modern" structural vocabulary: because of increased heat insulation requirements and the resultant greater insulation thicknesses, walls are getting thicker again. So the way edges are handled is once again topical.

Wall thickness

Wall thickness and reveal design, the light admitted, and solar energy use are all affected, with the same aperture area. In addition, building

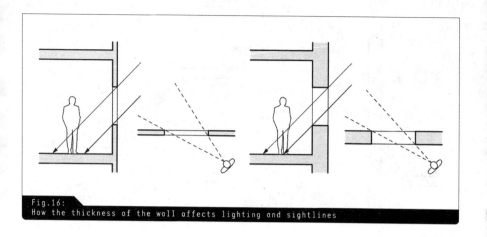

Fig.16:
How the thickness of the wall affects lighting and sightlines

type and construction method also affect sightlines from inside to outside, by expanding or restricting the visual links. › Figs 15 and 16

In non-loadbearing curtain facades, which have narrower wall superstructures and thus considerably less depth of shade, the dimensions of posts and rails extending into the room have to be taken into account, as well as the thickness of the glazing.

Geometrical design

The depths of the edges depend on the structure of the wall. The design is associated with the rebate on the elements closing the aperture (e.g. windows or fixed glazing).

The edges can be beveled, in order to increase the amount of light admitted in the case of small apertures, or to make them seem larger. These (window) incisions can be arranged diagonally facing outwards or inwards. The amount of daylight admitted can also be increased by painting the edges of the reveal a light, highly reflective color.

The edge of the aperture can be changed all the way round or individually, symmetrically or asymmetrically. Regardless of the material quality of the exterior wall, the beveled edges can be in exposed masonry or can be appropriately rendered, in concrete or natural stone.

Outward-facing beveling

Beveling facing outwards in the lintel area enlarges the proportion of zenith light admitted to the interior. A clearly inclined sill improves rainwater drainage and gives an enhanced sense of connection with the outside

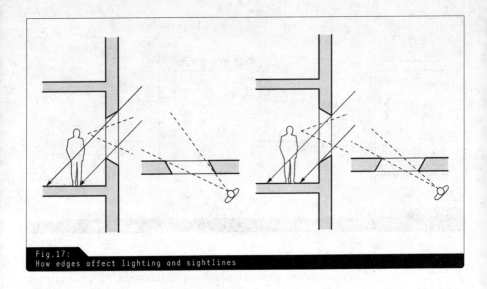

Fig.17:
How edges affect lighting and sightlines

world, especially in multi-story buildings; and a window breast can also act as a sightscreen. › Fig. 17

Inward-facing beveling

Reveals widened on the inside reduce the light density contrast between wall and aperture and thus prevent glare. The straight edge reinforces the silhouette effect. There is a relatively small edge between the opening, toplit on the room side, and the dark wall. In case of a cloudless sky and areas exposed to the sun, the beveling creates a medium-bright transitional zone. Depending on the reveal depth and angle, there is little or no glare.

Make the window frame a deep, splayed edge: about a foot wide and splayed at about 50 to 60 degrees to the plane of the window, so that the gentle gradient of daylight gives a smooth transition between the light of the window and the dark of the inner wall." (Christopher Alexander et al., *A Pattern Language,* Oxford University Press, New York 1977, p. 1055)

In contrast to historical examples, many contemporary buildings come up with a number of asymmetrical solutions. These create a special sense of closeness to local climatic conditions or features of the urban situation. › Figs 18 and 19

Care should be taken when finishing edges geometrically so that they relate correctly to the size of the aperture and its ratio to the wall area. Staggering or stepping individual structural elements or offsetting edges

Fig.18: One edge beveled

Fig.19: Two edges beveled

makes the wall look sculpture externally. This effect is further enhanced by the resultant play of light and shade, the different surface materials, and the varied colors. The depth of the edges can also be increased, but not reduced, by shifting other structural elements off the plane of the outer skin (jambs or extended reveal boards). > Chapter Aperture components, Elements

Even though there is little quantified material available about the interplay of aperture size, edge design, light admission or solar energy use, it opens up the aperture planning repertoire enormously on the design plane.

CONTROL DEVICES

Control devices are added to apertures for planned manual or mechanical control of climate and weather effects on the space. Structural devices like this can be used to regulate the degree of permeability and thus air quality, as well as the temperature and moisture levels between indoor

\\ Example:
Examples of geometrically accentuated aperture edges can be found in buildings by the Frankfurt architect Christoph Mäckler, and the Stuttgart practice Lederer Ragnarsdóttir Oei.

Fig. 20:
Shutters with movable slats

and outdoor space, according to the time of day or year, or can provide full or complete shade for radiation or transmission. Several such devices can be combined and used together according to the movement mechanism and the amount of strength required.

The search is also on, while aperture areas are being enlarged, for materials and structural elements that can be used to influence the extent of the desired permeability. Early versions of these devices were made of skins, fabric or paper. These were later followed by (wooden) shutters that could be turned or slid, or covered with semi-transparent organic materials. These simple wooden shutters have developed into a wide variety of movable elements as existing ones have been refined and new systems devised. › Fig. 20

Control devices can also be operated mechanically. Sensors make it possible to control the system automatically, in line with the weather. Here it makes sense for users to be able to override the system individually if they wish. Comfort levels and energy consumption can be optimized by combining various elements or principles.

Positioning

Positioning such control devices affects their functional context directly. This applies to their:

_ Position in relation to the opening (top/middle/bottom/one or more sides)

_ Position on the outer wall (on the outside, away from the opening/ on the outside/built into the window plane/inside)

Functional properties

Control devices differ in their technical and material design, and particularly in the way they are operated and the degree of variability; in this case the extent to which the opening is enlarged. In addition, the nature and direction of the movement have a part to play.

A wide variety of finishes are available in the familiar systems. There are three ways of looking at the main typological features:

_ Permeability
_ Movement (by the element)
_ Packed size

Permeability

The nature and extent of permeability for light, air and heat has a crucial effect on the qualities required of a control device. Intermediate positions can be set between open and closed. The desired degree of permeability must then be set (e.g. from fully open to a narrow slit).

Movement

Control devices can be categorized as movable or constructed to move. This can apply in terms of time: movable temporarily, but fixed, i.e. as secondary double glazing or a window shutter, or permanently movable like a folding shutter or roller blind.

In this book control devices are treated as permanently movable elements. There are also systems that do not move at all (e.g. switchable special glass, gasochromic or electrochromic glass). › Chapter Window components, Special glass

Control devices can be assembled from different components. They can move in different ways, and create different conditions and variable degrees of permeability.

Packed size

The packed size, i.e. change in relative dimensions, is crucial to using control devices. It can remain unchanged (as in a hinged shutter), reduced (as in a folding shutter or – even more clearly – in the case of slatted blinds, where the packed height is about 6 to 10% of the maximum area covered), and directly affects the way they are handled.

Nature and direction of movement

Types of movement are often a combination of movement principles. Combined with direction, they offer a variety of possibilities. Comfort can be regulated efficiently when it is possible to respond to the effects of

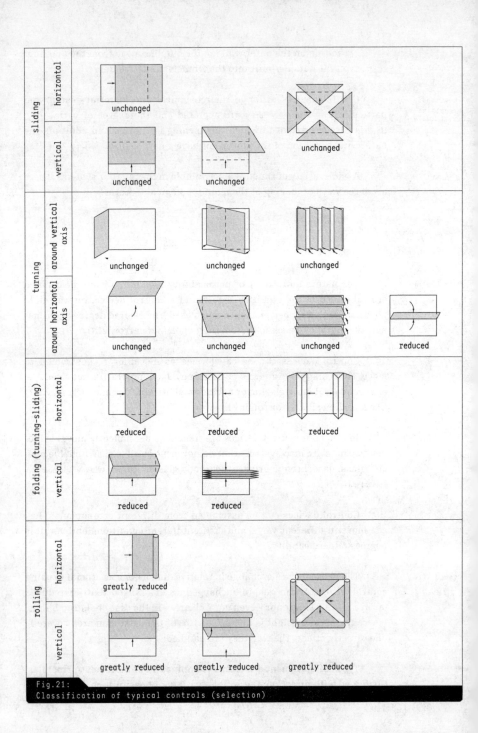

Fig.21: Classification of typical controls (selection)

light, sound and heat technology independently of each other. Movement can be

_ Around the vertical axis (rotation, roller shutters)
_ Around the horizontal axis (rotation, horizontal slat)
_ Complete change of place without changing the element (translation, sliding shutter)
_ Complete change of place and of the element (transformation, roller blind)

The kinds of movement can also be categorized in terms of operation and the amount of space needed with reference to the facade plane – usually outside/inside and top/bottom:

_ Turning: opening inwards/outwards
_ Folding: opening inwards/outwards
_ Sliding: horizontally (to the right/left/vertically (upwards, downwards) > Fig. 21

Construction types

The simplest control device for closing apertures is the shutter, which is commonly made of wood as the material is readily available, works well, and is easy to handle. There are also examples in natural stone and, since the 19th century, metal as well in many places. Shutters were used as an alternative to windows at first, and then in addition to them, from the 15th century onwards.

Movement and fixing types

Control devices can be categorized by movement and fixing type:

_ <u>Sliding shutters</u> (sliding horizontally), placed at the side of small apertures, set on tracks inside or outside > Fig. 22
_ <u>Dropped/raised shutters</u> (sliding vertically), set above or below the aperture, usually set into the wall structure

\\Example:
Window shutters can be used to show the device's increasing sophistication and increasingly complex construction. Starting with the opaque timber shutter, the panels then acquired translucent or transparent apertures to provide a minimum of light and view outside. In about 1700, the flat boards were replaced with fixed diagonal slats. They later became movable manually: a strip of wood or a metal bar could be used to adjust the amount of light permeating. There are early examples with the shutter divided into different functional areas, so that the light admitted, the view through, ventilation, and privacy screening can be adjusted independently.

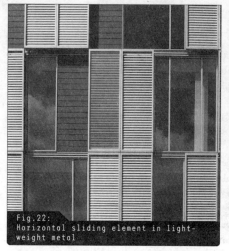
Fig. 22:
Horizontal sliding element in lightweight metal

Fig. 23:
Wooden slatted fold-and-turn shutters

- <u>Hinged shutters</u> (moved by turning), fixed on hinges to the side of the aperture; the variant is vertically slatted shutters › Fig. 20
- <u>Box shutters</u> (moved by folding and tilting), fixed on hinges above or below the aperture
- <u>Folding shutters</u> (moved by a combination of sliding and turning), fixed on side hinges › Figs 23 and 24
- <u>Roller shutters</u> (narrow slats, strung on strings or chains) or (membrane) roller blinds (moved by rolling), fastened above the aperture, partly let into the wall structure
- <u>Slatted or Venetian blinds</u> (narrow slats, fastened together on strings; moved by a combination of sliding and turning), fitted above the aperture, partly let into the wall structure › Fig. 25

Control devices are available in almost all the usual construction materials. If many combinations of individual components with different movement mechanisms are involved, care should be taken to ensure that these do not prevent each other from working properly.

Element sizes

The element size or cross section depends on the dimensions and movement of the control device. Hence upright formats are appropriate in terms of handling the effect the load makes on fittings and support structure for sliding and folding shutters, horizontal formats for box shutters. Sliding shutters can jam very easily if the proportions of the long and narrow sides are extreme. Care should be taken over the span width for linear systems such as slatted and roller blinds to prevent bending.

Fig. 24:
Opaque wooden fold-and-turn shutters

Fig. 25:
Lightweight metal Venetian blinds

Practical example: shading devices

It is important when choosing a control device to be familiar with the interplay between the effect of the weather and the principle on which the device works. When providing protection from the sun (shade for the aperture), it is necessary to respond to the different climatic conditions in each case, and the changing daily and seasonal position of the sun throughout the year.

Various principles can be identified by taking the provision of shade for a south-facing aperture as an example:

_ Complete, direct covering for the aperture: this blocks visual contact with the outside world, so artificial light will be needed in the room.
_ Semi-transparent structures (perforations, expanded metal, etc.): extensive shading allows some contact with the outside world and admits some light. › Fig. 26

Slatted structure

It is more effective to divide the area up with an accumulation of small elements in a slatted structure: the slats can be adjusted as the sun moves – and still provide the same amount of view outside. The best approach uses non-linked, separately controlled systems. This provides shade and a possible view outside on the sightline. Incident daylight can then independently be directed in the upper section of the window. These systems provide good protection from the sun, while allowing natural lighting and visual contact, which can be optimized by using semi-transparent slats. › Fig. 27

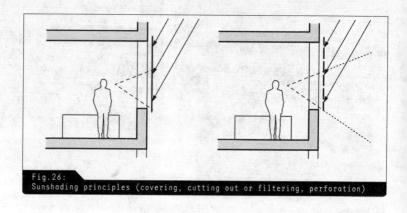

Fig. 26:
Sunshading principles (covering, cutting out or filtering, perforation)

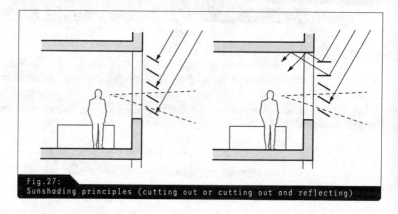

Fig. 27:
Sunshading principles (cutting out or cutting out and reflecting)

Positioning principles

Two positioning principles can be applied to slatted structures, based on the direction in which the window faces and the associated position of the sun:

_ <u>Horizontal slats</u> prevent sunshine at a steep angle on the south side from penetrating the space. The shallower the radiation angle (east and west side), the more comprehensive the shading will need to be.
_ If apertures face east and west, <u>vertical slats</u> are used to prevent transmission.

Shading is provided, but it is still possible to see out. › Figs 28–30

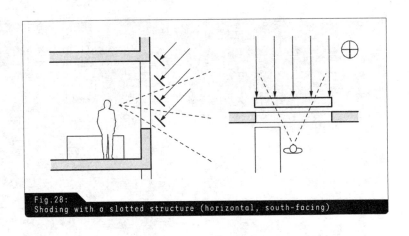

Fig.28:
Shading with a slatted structure (horizontal, south-facing)

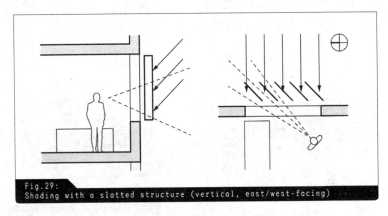

Fig.29:
Shading with a slatted structure (vertical, east/west-facing)

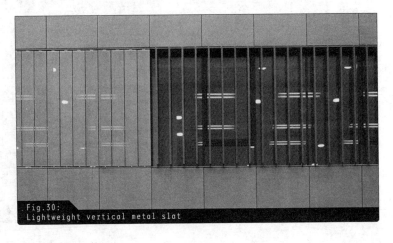

Fig.30:
Lightweight vertical metal slat

APERTURE COMPONENTS

ELEMENTS

Lintel

The topmost element of a wall opening is the lintel. It spans the opening, and transfers loads into the loadbearing wall cross sections at the side, the reveals. The simplest lintel is a beam that can carry a bending load (e.g. wood, reinforced concrete, steel). The aperture size is limited by length limits specific to the material and the maximum admissible deflection of the beam. Pressure-loaded lintel constructions are traditionally used in masonry building in particular (e.g. horizontal arches). The supports have to absorb both vertical and horizontal loads from the arch structure.

Reinforced concrete lintels can be cast on site at the same time as the ceiling, or supplied as a prefabricated lintel. The range of prefabricated lintels includes reinforced concrete units and brick shells with concrete-clad steel reinforcement, U-shell prefabricated lintels, and roller blind lintels.

Sunshading systems

For particular building orientations and exterior wall concepts a sunshading system (e.g. roller blinds, curtaining) may have to be built into the lintel or attached to the outside. Integrating them into the lintel area can affect shell geometry and thus the protective function of the external wall.

Apertures can also be closed flush with the ceiling, without a lintel. This makes the interior look more generous and admits more daylight. Where necessary, the edge of the ceiling slab is reinforced to span the opening.

Window breast

The bottom of an aperture can take the form of a window breast. Solid masonry sections or fixed window element must be high enough to prevent people from falling out. According to the position of the window in the outside wall, a horizontal internal covering, a windowsill, should be provided; this is particularly important outside.

Radiators are often fitted inside the space in the window breast area; these reduce the amount of cold air flowing down from the glass panes,

> \\ Note:
> For more information on static systems, loadbearing behavior and supports see *Basics Loadbearing Systems* by Alfred Meistermann, Birkhäuser Publishers, 2007.

among other things. The customary radiator niches must not significantly detract from the statical efficiency of the wall cross section and the wall's heat permeation resistance.

As a rule, French windows do not have a breast. The bottom of the door must be adequately sealed, which is usually achieved by adding a threshold. It is possible to provide threshold-free access to the outdoor space or a balcony by using grids or gutters. The balcony or terrace must be fitted with protection against falling if it is more than one meter above the ground.

Rebate

The rebate is the shallow connection between the window frame and the shell of the building. We distinguish between › Fig. 31

(a) Window with internal rebate › Fig. 33
(b) Window without rebate
(c) Window with external rebate.

Independently of this, the window can be placed on various planes.
› Fig. 32

(d) Flush with the interior
(e) Centrally in the reveal
(f) Flush with the exterior

The insulation layer must always come right up to the window. There are usually rebates on three sides, in each reveal and in the lintel area. Windows with a reveal – central positioning is most common, with an interior rebate – make it possible to create a chicane between the shell and the window frame, while for windows without rebate the requirements for a joint (sealing, insulating, fixing) must be fulfilled within the depth of the frame.

\\ Example:
Maximum acceptable drop height
If windows open, the required window breast height must be met with solid window breasts or fixed glazing elements to prevent people from falling out. For opening French windows there must be additional protection against falling inside or outside, depending on the direction of opening, e.g. in the form of a railing.
Breast or railing heights depend on the height and purpose of the building. Solid window breasts should be ≥ 80–90 cm high up to the 12 m drop height (except on the ground floor), above 90–110 cm. Window breasts may be lower if an additional guardrail is provided.

40

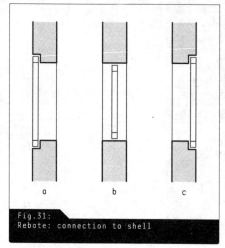

Fig.31:
Rebate: connection to shell

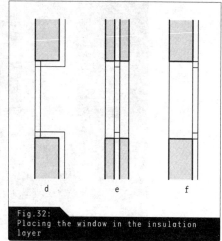

Fig.32:
Placing the window in the insulation layer

External rebate

Reveals with exterior rebates are traditionally found in particularly stormy areas (e.g. North Sea Coast). Wind pressure means that the outward-opening windows – and the window as a whole – benefit from being pressed against the seals and rebates.

Today, the fitting of the windows is the key criterion for the type of rebate. For multi-story buildings in particular, it is possible to fit windows from the outside and replace window elements from outside if they are damaged, using scaffolding or lifting devices. If elements are large or particularly heavy it can make sense to create an exterior rebate to assist lifting devices (crane).

If a window is mounted flush with the exterior, frames, glass and connecting joints are particularly exposed. Joints have to be realized with

\\ Note:
A chicane is the term for a geometrically shaped structural element, e.g. with a rebate or groove, a built-in obstacle in the wall structure to prevent water penetrating directly.

\\ Important:
Insulation layer – rebate
Windows should be placed within the facade insulation layer, to avoid heat bridges. The rebate also prevents precipitation from penetrating, through the principles of chicane and overlapping.

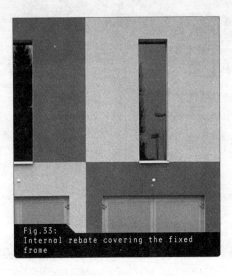

Fig.33: Internal rebate covering the fixed frame

particular care. In terms of building physics, it is difficult to fit a window flush with the exterior even with an external rebate (peripheral joints as chicane), as the dew point is shifted a long way outwards in the window reveal area and heat bridges can be formed. This requires interior insulation of reveal and lintel where necessary. On the inside, it is possible to realize a flush transition between the reveal and the glazing in the case of fixed-glazed windows with an external rebate.

Internal rebate In the case of an internal rebate, the window can usually be fitted from the interior without any additional effort and replaced in case of damage. The frame is pressed against the rebate from the inside and is set back in the outside wall by the depth of the rebate. The breadth of view of the window frame in the lintel area and at the reveals depends on overlapping, and can be reduced to the opening frame profile. › Fig. 33

Internal rebates offer a higher degree of safety, as the rebate overlaps the frame and the rebate joint can be realized as a chicane. The window can be fitted flush with the interior wall. Similarly to the exterior flush fitting, counter-measures must be taken against position-related heat bridges (e.g. interior insulation or insulation for the reveal).

Without rebate If the wall aperture is realized without a rebate, this simplifies work on the shell in the first place. But greater demands are made on the joints, which have to be both wind- and vaporproof, and also free of heat bridges, but – unlike fitting approaches with rebates – only the depth of the window profile is available for this.

As long as the window is fitted within the insulation layer, or the insulation is taken to the window in the reveal area, the position of the window within the thickness of the wall can be chosen freely (flush outside, flush inside, central). The window frame is visible from both the inside and the outside over almost its full width. The bottom of a window is usually finished without a rebate, in order to meet the functional and geometrical demands of the jointing process.

Vertical loads from the window element are transferred into the loadbearing wall cross sections. Precipitation water collects along the windows and has to be deflected outwards at the bottom, e.g. by providing a sill. The wall cross section – for additional-leaf constructions the insulation in particular – is protected against damp. In the case of rear ventilated wall constructions the bottom of the window must also allow free ventilation. > Fig. 57, page 73

On the inside, the coverings (for windows with breasts) and the floor fittings (for French windows) must be attached to the bottom frame section. In addition, waterproofing seals must be fitted at the bottom of French windows in conformity with directives (e.g. flat roof directives).

BASIC CONSTRUCTIONS

Openings in the vertical component of the envelope system – the facade – are usually closed by doors and windows. These combine opening and lighting functions with the protective function of the wall. Window frames standardized by manufacturing and profile configuration are fitted into different wall constructions.

\\Note:
French windows
The waterproofing should be attached 15 cm above the surface covering or gravel to prevent weather-related water penetration above the door or French window threshold, for example if drains are blocked or precipitation water freezes. Favorable local conditions can make it possible to reduce the attachment height, provided that water can run off without difficulty at all times (e.g. by placing a gutter outside the window aperture). But the minimum attachment height of 5 cm should adhered to even then.

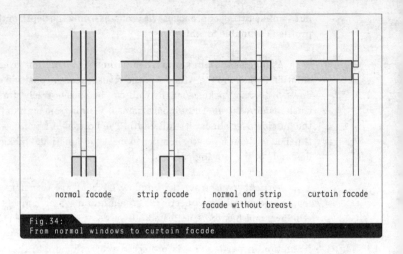

Fig. 34:
From normal windows to curtain facade

normal facade | strip facade | normal and strip facade without breast | curtain facade

Normal windows

In the normal form for windows, the openings finish at the top with a lintel, at the sides with reveals and at the bottom a window breast between them and the wall cross sections. Such windows can be combined to form continuous or strip windows. Here the side connections within the strip of windows consists of window elements (visible frames), with tops and bottoms of the normal kind.

Curtain wall facades

In curtain or window wall facades, element joints must be created at the top and the bottom. This leads to larger facade elements consisting of different window units. Vertical posts and horizontal rails articulate and subdivide multi-member window elements. › Figs 34 and 35

Aperture size

The desired aperture size directly influences the construction and material quality of the facade. Wall construction and formats define each other. Tall, narrow apertures are appropriate in masonry walls. Wide openings need high lintels and reinforcement in the reveal area. Cast reinforced concrete walls tend to be statically overdefined (the loadbearing capacity of the wall can be additionally increased with reinforcement that is invisible from the outside). This means that the openings can be enlarged correspondingly. In skeleton construction the opening can be taken as far as the columns and beams (walls and ceilings in crosswall construction), the dimensions of which are determined by statical requirements. › Fig. 36

If wide spans have to be tackled or the fixed glazing is high, the wide window sections can be replaced with narrower clamping strip sections. Opening windows have to be fitted with their own frames in this case. All

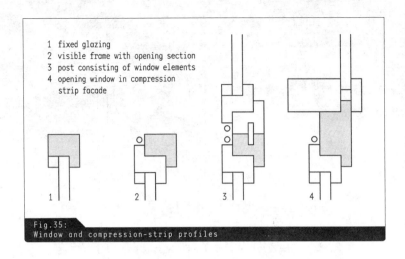

Fig.35:
Window and compression-strip profiles

1 fixed glazing
2 visible frame with opening section
3 post consisting of window elements
4 opening window in compression strip facade

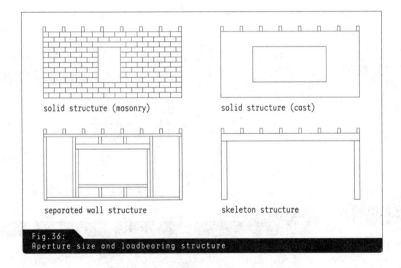

solid structure (masonry)

solid structure (cast)

separated wall structure

skeleton structure

Fig.36:
Aperture size and loadbearing structure

the major manufacturers offer window and clamping strip sections in various materials and a wide range of dimensions, some of them standardized.
> Chapter Window components, Frames

Connection to wall

The way a window is connected to a wall requires different approaches at the sides, top and bottom. The connection area is most heavily loaded in the window breast area, and in the lintel area the need to build

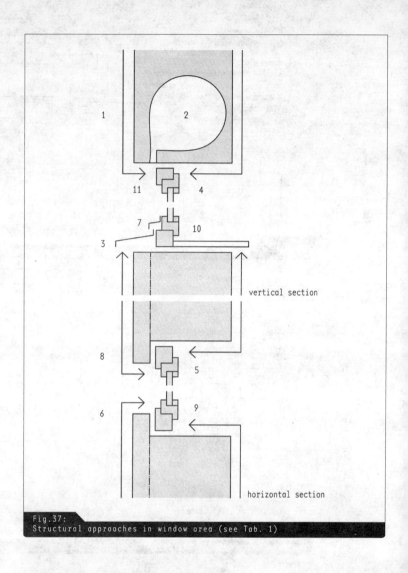

Fig. 37:
Structural approaches in window area (see Tab. 1)

in roller blinds and sun protection systems makes clear geometrical solutions difficult. > Fig. 37 and Tab. 1

General structural principles such as direct load dispersion, continuous insulation, adequate tolerances, seals between different structural elements, chicanes, and overlaps are addressed in specific strategies for constructing the window joints.

Tab.1:
Aperture planning principles (see Fig. 37)

		side	top	bottom
1	Load dispersed via opening, lintel		x	
2	Integrating control devices	x	x	(x)
3	Exterior drip edge		x	x
4	Tolerance, seal between frame and wall	x	x	x
5	Fixing the window (tension-free)	x	(x)	
6	Taking the exterior cladding up to the window element	x	(x)	(x)
7	Heavy rain seal outside	x		x
8	Rebate for the window element	(x)	(x)	
9	Taking the interior cladding up to the window element	(x)	(x)	
10	Change of material for the reveal (sill, framing)	(x)	(x)	x
11	Insulation in the reveal area	(x)	(x)	(x)

x always
(x) possible

WINDOW COMPONENTS

OPENING TYPES

Windows that can be opened operate in a number of different ways.
> Chapter Frames and Figs 38–40

Turn, tilt, tilt-turn window

The turn window can be opened inwards or outwards on a vertical axis, and the tilt-turn window can also be tilted inwards on the lower horizontal axis. For ease of use, turn, tilt and tilt-turn windows usually open inwards. One variant is the turn-fold window, which can open inwards out outwards. > Fig. 41 It is also important to establish whether the window is hinged on the right or the left.

Top-hung or flap window

Flap windows open outwards. The turning axis is placed horizontally at the top, to prevent rain coming in when the flap is open. > Fig. 42

The rebates for opening windows and frames are stepped on the outside, corresponding to the direction of opening. > Fig. 43 The upper joint between the frame and the opening window is susceptible to heavy rain. A weather bar is usually fitted so that precipitation water can run off and not down the facade.

Sliding, lift-and-slide windows

Horizontal sliding windows slide sideways and usually open on the inside of the fixed element for reasons of sealing tightness. Lift-and-slide windows are raised vertically before sliding, which makes them easier to operate.

Vertical sliding window often open upwards in the case of two-part windows, but also downwards in the case of large windows without breasts. > Fig. 43 Here the sliding sections can be lowered into the floor (e.g. into the basement story), so that they can be stepped out of at ground level.

\\ Note:
The way windows open should be shown in the working plan views on a scale of 1:15: turning movements are drawn in as unbroken lines in exterior views (opening outwards), or dashed lines (opening inwards), and sliding movements as arrows in the direction of opening. When dimensioning the opening on ground plans the height of the opening is entered as well as width, with the width figure placed above the dimension line and the height figure below it.

Further information on plan presentation can be found in *Basics Technical Drawing* by Bert Bielefeld and Isabella Skiba, Birkhäuser Publishers, 2007.

The floor-to-lintel height of a window or door aperture is placed under the dimension line in the working plan

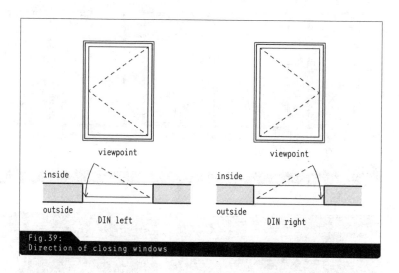

Fig.38:
Dimensioning windows in the working plan

Fig.39:
Direction of closing windows

\\ Note:
Closing direction for windows according to DIN EN 12519:
"Left": window opens on the view side with hinges on the left-hand side; closes counter-clockwise
"Right": window opens on the view side with hinges on the right-hand side; closes clockwise.

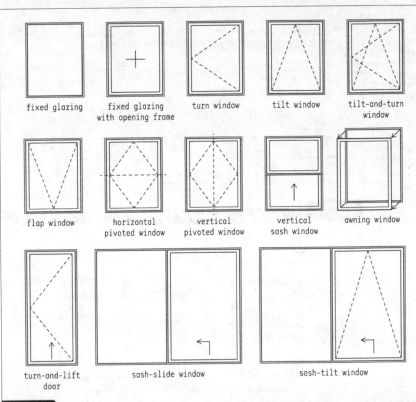

Fig.40:
Aperture types as shown on the working plan

Fig.41:
Fold-turn window (metal)

Fig.42:
Flap window (aluminum)

Abb. 43:
Vertical sash window (wood)

Lift-slide-tilt windows

Lift-slide-tilt windows can also be tilted inwards around the horizontal axis, which greatly increases the complexity of the fittings, and the amount of space they occupy.

Horizontal pivoted window

A horizontal pivoted window turns horizontally around its central axis, i.e. the outside of the pane can be turned into the interior, which makes cleaning the glass easier, among other things. The centrally placed pivotal hinge means that the profiling and the sealing rebates are offset. Only half the window aperture is accessible as a free cross section at any time. This is relevant if the window is intended to be part of an escape route. Windows of this type were commonly used in the 1950s and 1960s, but are quite rarely fitted today, mainly because of sealing problems around the pivot.

Vertical pivoted window

The vertical pivoted window uses the same principle, and has raised the same problems. The opening window turns around the central axis in a vertical direction.

Slotted window

A series of small flaps, or vertically or horizontally pivoted slats, is called a slatted window. If each of the slats has its own frame, the proportion of frame to the glazed area is relatively high. The dimensions of the individual slats fall between at least 200 × 100 mm, or a maximum of 2000 × 400 mm. Element sizes run from at least 300 × 150 mm to a maximum of 2000 x 3000 mm.

Awning window

Awning windows are pushed in front of the facade plane by a scissor device, thus providing excellent ventilation: cold air can flow in from

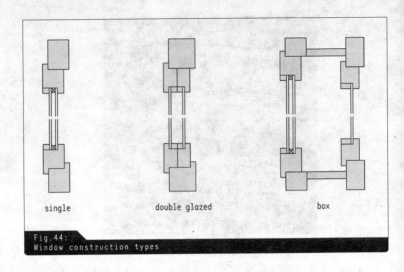

Fig. 44:
Window construction types

below, and warm air escapes at the top. As none of the sides is firmly fixed to the frame, the extending mechanism must exert completely even pressure.

CONSTRUCTION TYPES › Fig. 44

Single windows

Single windows are the norm nowadays, as thermopane glazing can provide good heat insulation. So the U_w value of a window with thermopane glazing is about 3.0 W/m²K and with thermal insulation glazing including frame about 1.3 W/m²K.

Double-glazed windows

Before insulating glass was developed, windows were single (U_w value approx. 4.8 W/m²K). Double-glazed windows were much in evidence to improve heat and sound insulation. Two single frames were combined, i.e. opened together as well. If the panes are 40 to 70 mm apart, a slightly better insulation value is achieved in relation to single windows with insulating glass (U_w value approx. 3 W/m²K).

\\Note:
Overall heat transfer coefficient
Window structures have to be evaluated as a complete system in terms of their heat insulation capacity. Windows are allotted a U_w value. The U_w value (formerly k value) states the energetic quality of the entire structural element. The lower the U_w value, the less warmth is lost through the window surface. The U_w value is made up of the individual heat insulation values of the glass (U_g value), the frame (U_f value), and the linear heat transfer coefficients of the glass edge (psi value), taking the area proportions into consideration.

Fig. 45:
Box windows (steel)

The sections could be opened up for cleaning. Double-glazed windows are usually supplied as turn, tilt-turn and tilt windows.

Box windows

Old buildings often still have double-glazed windows made up of two simple panes mounted together. These are further apart than the frames in modern double-glazed windows, and are not combined structurally. Now the inner section usually contains insulating glass. The airspace between the panes increases heat and sound insulation. If the two windows placed one behind the other are made into a unit with a continuous lining, usually made of timber, the unit is known as a box window. Both types are usually supplied as turn windows. > Fig. 45

Box windows are often divided by bars, a typical 19th-century window type. The high proportion of frame, the bars, and the offset positioning of the frames mean that a lot of daylight is lost. The box window is a forerunner of today's additional-leaf glass facades. Because they are complex and expensive to manufacture, box windows are mainly used when there are stringent demands on sound insulation or for refurbishment in listed buildings.

Frameless window

In modern windows, it is the frame, not the insulating glass unit, that is the weak point in terms of heat transfer. One possibility for optimizing this lies in reducing the proportion of frame. The chief characteristic of frameless windows is that they have no visible frame or glass retention devices (point holders, clamp profiles) on one side. The glazing consists of stepped-edge glass: the outer pane protrudes over the edge of the inner pane and is glued to the supporting frame. The adhesive can be made out as a continuous black margin behind the outer pane of glass.

Frameless windows can be supplied in most of the opening methods mentioned. However, the glued joint is more delicate and needs more maintenance than a clamp profile.

Fig. 46:
Components of a window

FRAMES

Fixed frame

This type of frame is built into the wall structure and the frames of the opening sections are attached to it, or fixed glazing is mounted in it.
› Fig. 46

- (1a) Frame wood, top section of the fixed frame
- (1b) Lower section of the fixed frame (not shown in Fig. 46)
- (1c) Side section of the fixed frame
- (2) Post (window post), dividing the fixed frame vertically (not shown in Fig. 46)
- (3) Rail (crossbar), dividing the fixed frame horizontally

When fitting the fixed frame, there should be adequate tolerance with the wall opening, to avoid tensions and absorb movement in both the structural element and the building. Connections with the building must be slightly elastic or free to slide. › Chapter Fitting structural elements together, Seals

Casement frame

The casement frame is the part of the window that is attached to the fixed frame and can be opened. › Fig. 46

- (4) Casement frame
- (4a) Casement timber, upper section of the casement frame
- (4b) Casement timber, lower section of the casement frame (weather bar)
- (4c) Casement timber, vertical section of the casement frame
- (5) Bar, to divide the casement frame horizontally
- (6) Glazing

French casement

Windows with two opening sections can be divided by a post, the window post, i.e. a fixed elements that is part of the fixed frame. Alternatively, a rising centerpiece, the astragal, can be used as part of the opening section of the window, in order to cover the joint on this section and the sealing. This is known as a French casement.

> \\ Important:
> Permissible tolerances limit deviation from nominal dimensions, from the size, shape, and position of a structural element in the building. The term "tolerance" also includes intended deviations from the theoretical nominal dimensions. This enables compensation for imprecision that is inherent in materials or specific to the manufacturer.

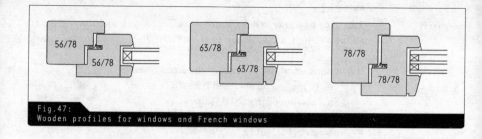

Fig. 47:
Wooden profiles for windows and French windows

MATERIALS – SYSTEM WINDOWS
Timber windows

The chief advantages of timber windows are its good heat insulation capacity (comparative thermal conductivity: spruce: $\lambda = 0.11$ W/mK; aluminum alloy: $\lambda = 209$ W/mK), ease of working, and the sustainability of the material. There is no need for elaborate and expensive manufacturing processes or complicated thermal separation of the profile cross sections.

Coniferous timber (pine, spruce, lark, Douglas fir, fir) are mainly used for window construction, and deciduous timber more rarely (oak, robinia). The use of tropical deciduous timber (Meranti, mahogany, Kambala) is declining. The surface of the frame is subject to constant weathering by UV rays, drying, rain, etc. The constant change in the timber's moisture content that this brings about (25–50%) encourages damaging attacks by putrefying fungi, mold or insects. A timber that is suitable for window construction must thus be adequately robust, have a low moisture absorption capacity and show natural resistance. › Fig. 49

Natural timber windows can be additionally protected by surface treatment. Priming is a preventive measure against timber-discoloring fungi, and impregnation prevents moisture-induced rot. A distinction is made between varnishing and covering coats of paint. Please note here: the darker the color, the more the timber will heat up. Beveled or rounded profile edges ensure that the paint will bond with the timber profile more durably.

Structural timber protection is essential when constructing timber windows. Standing water must be avoided, and precipitation water must be able to run off timber profiles and surfaces, i.e. there should no horizontal surfaces.

Profiles

Timber profile cross sections are available in standard sizes and gradations (standard profiles). › Fig. 47 Brief definitions give the profile depth

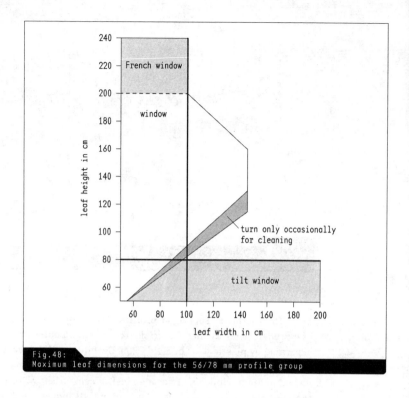

Fig. 48:
Maximum leaf dimensions for the 56/78 mm profile group

and profile length in millimeters. U_f values of 1 W/(m²k) can be achieved with window frames in solid wood (e.g. 3-ply glued wooden frames).

Element sizes

The profile size affects the maximum leaf size, both relating to the nature of the opening. The leaf of a tilt window should for example not be higher than 80 cm with a profile cross section of 56/78 mm; the leaf of a French window (turn door) should not be wider than 1 m because of the dead weight and the load on the hinges. › Fig. 48

Metal windows

Metals are very good heat conductors. Steel has 250 times the thermal conductivity of wood, and aluminum has 4 times the thermal conductivity of steel. So metal profiles must be separated thermally in window construction.

There is a wide range of products available for both aluminum and steel windows. The profile dimensions differ considerably. Unlike timber windows, there are no uniform standard profiles fixed.

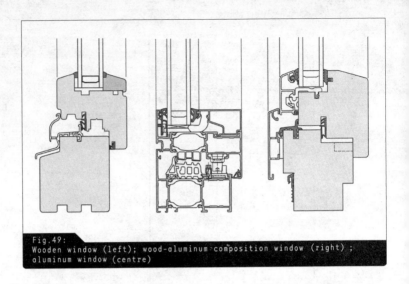

Fig. 49:
Wooden window (left); wood-aluminum composition window (right); aluminum window (centre)

Aluminum window

The thermally separated profiles of aluminum windows are made of extruded semi-finished materials. The inner and outer profiles are also separated by plastic pads or rigid foam. Despite the high investment costs (because of the amount of primary energy used in working the raw materials), aluminum windows are economical in comparison with windows of the same size in wood or plastic. Aluminum lasts longer, and is considered particularly easy to maintain and care for. It has low profile tolerances, i.e. the frames are very precisely dimensioned; it is very workable, and very light in weight. > Fig. 49

Points where aluminum windows connect with the rest of the building have to be able to absorb larger temperature-related length changes than timber or steel windows. (Aluminum expands in length by 1.5 mm at a temperature difference of 60 K.) Adequately dimensioned expansion joints between the fixed frame and the building, and inside large window and facade elements must be planned in.

As a rule, the surface is coated, as untreated aluminum oxidizes irregularly and stains. We distinguish between mechanical surface treatments such as smoothing, brushing or polishing, and electronic processes such as anodizing, which produces an even layer of oxide. Aluminum profiles can also be stove-enameled or powder-coated. Powder-coating is achieved by applying the coating material with powder guns and baking it in an oven at object temperatures of about 180 °C.

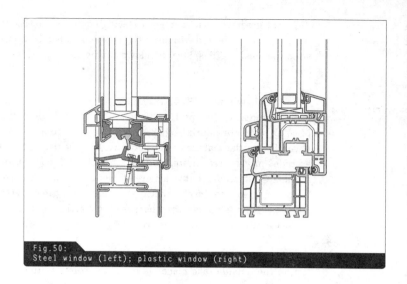

Fig. 50:
Steel window (left); plastic window (right)

Double-glazed windows (timber-aluminum)

Timber and aluminum double-glazed windows combine the positive properties of both materials effectively. The timber frame with its low thermal conductivity is placed on the inside as the loadbearing structure and the weather-resistant aluminum cladding on the outside as trim and weatherproofing.

Sliding joints should be used to compensate for the thermal expansion differential between wood and aluminum. Condensation must not be allowed to form in the contact zone.

Various construction types can be used for timber-aluminum windows. The external aluminum cladding can be fitted to a standard timber window subsequently. In other systems, the aluminum profiles form the outer glass rebate plane and integrate the rebate seals. Such systems achieve U_f values of 1.3–1.5 W/m²K. Multi-layer timber frames and an efficient insulating central layer (e.g. in polyurethane foam) achieve U_f values of 0.5–0.8 W/m²K.

Steel windows

Steel window frames can be made simply from hot-rolled T or L steel profiles, or as special profiles. Hollow sections are now most commonest, cold-rolled from high-quality strip steel. Steel windows have high bending and torsional rigidity and are more robust than aluminum windows. The frames consist of thermally separated profiles. › Fig. 50

The greatest disadvantage of steel windows is the risk of corrosion, which can be countered with protective paint or galvanization, or by using stainless steel profiles. Steel windows can be stove-enameled or powder-coated.

Plastic windows (PVC, GRP)

PVC profiles

Extruded hollow profiles in rigid PVC (polyvinylchloride) are supplied as single or multiple-chamber systems. Multiple-chamber systems have a higher thermal insulation capacity. To make the frame, preprepared extruded profiles are mitered and usually welded. The PVC profile has low thermal conductivity, but is also relatively low in strength, so metal sections are fitted into the chamber profile to provide greater stability. Thermal separation is not required because the metal parts are in the middle of the overall cross section. › Fig. 50

Condensation drainage and the required vapor diffusion compensation to the outside take place via outlet apertures in the front chamber. Most PVC windows are supplied in white. Plastic profiles can also be dyed or coated, but not painted.

For dark colors in particular, warming from solar radiation causes greater longitudinal expansion, because of PVC's high thermal expansion coefficient. The effect of light can also bring about changes in the color shade.

GRP profiles

To avoid heat loss via the window frame, i.e. to improve the U_w value of the window, GRP profiles (glass-fiber reinforced plastics) are also used as frame profiles. GRP has low thermal conductivity and is characterized by high strength and rigidity, so no further reinforcement is required. GRP profiles can be combined with aluminum profiles.

GLAZING SYSTEMS

The glazing system is a combination of glass, the glazing rebate and the sealing into the frame. When planning a window construction, the properties required from the glazing system must therefore be considered along with the material and construction of the frame. This applies to the thermal behavior of the glass and the production-dependent pane sizes, and where applicable any special functions for the glazing (heat and sound insulation, fire prevention), the bedding of the glass, and the sealing processes. › Fig. 51

In addition to the thermal transfer coefficient of the window (U_w value), which is expressed in W/m²K, the possible energy gain through the glazing is important for energetic evaluation of a window. Energy gain is defined as energy transmission efficiency or solar heat gain (g value). The g value is calculated from direct solar energy transmission and secondary inward heat transmission (emissions from long-wave radiation and convection). The g value is given for the glazing used in values between 0 and 1 or between 0% and 100%. The higher the value, the more energy is entering the room.

Panes of glass are supplied in a wide range of qualities with specific properties.

Single-layer glass

Float glass

Glass manufacturers Pilkington presented a new production process for flat glass in the late 1950s, the float process. Liquid glass from a furnace runs into a bath of tin at a temperature of 1100 °C. The lighter molten glass spreads out over the molten tin in a ribbon, with two parallel bounding surfaces. Float glass panes are usually between a maximum of 6 and 7 m long, according to production, and between 1.5 and 19 mm thick.

The width is determined by various factors, including the maximum dimensions for transport. If the glass is to be transported on a low loader, the maximum height for passing under bridges of 4 m, the height of overhead traffic light equipment and the turning circle of the vehicles are the restricting parameters. Subtracting the height of the vehicle, this means glass widths of something over 3 m.

\\ Important:
Physical properties of float glass, TSG and HSG:
Specific weight: 2500 kg/m^3
Elasticity module: 70,000–75,000 N/mm^2

Thermal properties:
Thermal conductivity: 0.8–1.0 W/(mK)
U_w value: < 5.8 W/(m^{2}K)
Transformation temperature: 520–550 °C

Acoustic and optical properties
(for thicknesses of 3–19 mm):
Assessed sound reduction index: 22–38 dB
Luminosity transmittance (Lt): 0.72–0.88
Radiation transmittance: 0.48–0.83

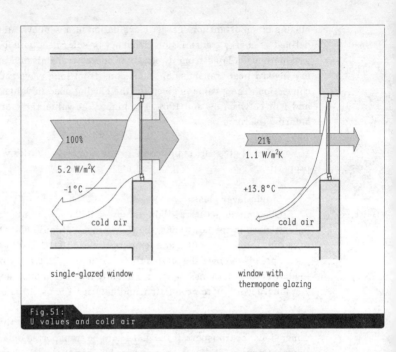

Fig.51:
U values and cold air

Tab.2:
Examples of U_w values for windows (2004)

	U_w value W/m²K
Single-glazed window	4.8
Window with double thermopane glass (details in mm) (4 / 12 air / 4)	3.0
Double thermopane glass (4 / 12 air / 4)	2.7
Triple thermopane glass (4 / 10 air / 4 / 10 air / 4)	2.2
Triple thermopane glass (4 / 8 gas / 4 / 8 gas / 4)	1.7
Double thermopane glass precious metal coated (4 / 20 gas / 4)	1.3
Triple thermopane glass precious metal coated (4 / 10 gas / 4 / 10 gas / 4)	0.9
Double windows/box windows	2.3
Special thermopane glass	0.4
Forced entry resistant glass	1.6

Fig. 52:
Glass panels, transparent, translucent

Varying from manufacturer to manufacturer, panes of float glass are manufactured 3.2 m wide and a maximum of 7 m long. Float glass can be colored in the production process. The natural green tinge of float glass can be reduced by a special selection of raw materials, producing "white glass" or "extra-white glass". Surface treatment like rubbing, etching or sandblasting makes the originally transparent float glass translucent. Float glass fragments into large, sharp pieces. It can be processed further to produce a number of products. › Fig. 52

Cast glass

When making cast glass, the liquid glass melt is passed between several pairs of rollers, creating one smooth and one structured or two structured surfaces on the glass. The surface structure makes the glass translucent, the light is scattered. The fracture pattern is the same as for float glass. Figured glass (e.g. U-glass) can be made by the casting process.

Wired glass

The hot glass mass has a wire net, welded at the intersection points and usually in stainless steel, pressed into it by a roller during the manufacturing process. This wire grid holds the parts of the pane together if the glass breaks, and prevents injuries. As a rule, wired glass (local regulations must be consulted here) can be used for overhead glazing. Water penetrating at the edges can lead to corrosion or ice according to the wire inlay, and cause flaking in the glass. Hence the edges of the glass should be sealed or protected by frames.

Thermally treated glass

Tempered safety glass

Heating the pane to the transformation temperature and immediate cooling (e.g. by blowing cold air over the glass) creates additional compressive stress in the glass, which prestresses it.

This change to the structure of the glass raises the flexural breaking resistance of the glass and its thermal shock resistance. If the glass is badly damaged, the result is a tight network of tiny glass crumbs, mainly with blunt edges. Tempered safety glass cannot be processed mechanically after the thermal treatment.

Heat strengthened glass

Heat strengthened (partially tempered) glass has a surface tension just large enough so that in case of damage it will break from edge to edge only. Heat strengthened glass (HSG) is not safety glass. It is made similarly to fully tempered glass (TSG), but the air-blowing process is slower, so that the compressive stress building up on the surface is lower than for TSG. Thus, HSG is less likely to break spontaneously than TSG. HSG has considerably greater flexural breaking resistance than float glass and much greater thermal shock resistance.

Multi-layered glass

Laminated glass

Laminated glass consists of at least two panes and an intermediate layer of foil or casting resin. It does not comply with any safety requirements, however.

Laminated safety glass (LSG)

The version that does meet safety requirements consists of at least two panes. So it is described as LSG of float + TSG, or LSG of double HSG, etc. The panes are fixed together by an intermediate layer of casting resin or sheet PVC foil (polyvinyl butyral). The sheets used are elastic and highly resistant to tearing, and they can be colored or calendared. They are inserted between the panes and welded to the glass under heat and pressure. The adhesive properties of PVC sheeting can be impaired if it is permanently wet. Hence the glass, for example as part of an insulated glass fitting, must always be fitted in an air-free rebate clearance to enable glass rebate ventilation. In case of breakage, the glass splinters from the

> \\ Important:
> Tempered safety glass (TSG) cannot be processed mechanically after thermal treatment, so any drilling or cutting to size must be decided upon and carried out before the tempering.

attached panes with be kept out of the intermediate layer, so that the glass still meets safety requirements in its damaged state.

Additional-leaf glass

Insulating glass

Insulating glass units usually consist of at least two panes of glass kept apart by a peripheral joint. The gap between the panes contains dried air or an inert gas to improve thermal or sound insulation. The peripheral connection is essential for impermeability and is usually made using a space, a primary sealant (e.g. butyl), and a secondary sealant (polysulfide). The spacer with its butyl strips is glued to the clean pane, and the second pane is then put in position and sealed with polysulfide.

The spacer has a cavity on the side of the gap between the panes and filled with a desiccating agent. A typical structure for an insulation glass unit is e.g. 6 mm float glass inside, 12 mm gap between the panes, and 6 mm float glass on the outside. › Fig. 53 and Tab. 3

Special glass

Thermopane glass

For thermopane glass, the thermal insulation properties of the insulating glass are enhanced by adding an inert gas filling to the gap between the panes and/or using three rather than the usual two panes (U_g value 0.5 Wm²K). The gap between the panes is increased by 10 to 16 cm according to filling.

Inert gases are use to fill the gap between the panes as they are poor heat conductors (krypton produces the best U_g value (0.5 Wm²K) with a gap of 2 × 12 mm). At the same time, panes of insulating glass are given a color-neutral coating of a precious metal to reduce thermal conductivity. This is applied to the outside of the inner glass pane.

\\ Note:
Manufacturers' standard insulation glass unit sizes
According to manufacturer, the maximum sizes vary between 420 × 260 cm and 720 × 320 cm, with a gross weight of up to 3.5 t. The dimensions of an insulating glass unit are based on the maximum area that can be manufactured. TSG with a thickness of 6–10 mm makes areas of up to 10.9 m² possible. Minimum dimensions for insulating glass units are 24 × 24 cm for float glass combinations, and 20 × 30 cm for TSG combinations. Coating processes are needed according to the properties required (thermal insulation, heat absorption, sound insulation), which can produce different dimensions.

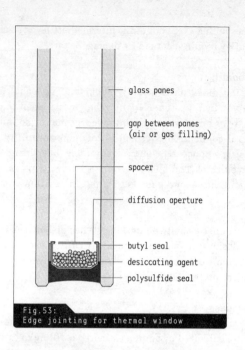

Fig.53:
Edge jointing for thermal window

Tab.3:
Building physics specifications for glazing types (as of March 2006)

	U_g value W/m²K	g value %	Lt value %
Single pane glazing (Pilkington Optifloat clear)	5.8	85	90
Double insulation glass with air gap 10–16 mm	3.0	77	80
Double insulation glass 4/16/4 argon, coated (Pilkington Optitherm S3)	1.1	60	80
Solar glass 6/16/6 with vapor coating (flat glass Infrastop Brilliant)	1.1	33	49
Triple insulation glass with 4/12/4/12/4 krypton, coated (Pilkington Optitherm S3)	0.7	50	72
Glass bricks	3.2	60	75
Single profile construction, (flange width) 60 mm (Pilkington Profilit)	5.7	79	86

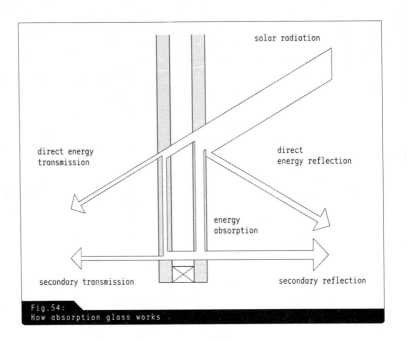

Fig. 54:
How absorption glass works

Heat-absorbing glass

Heat-absorbing glass is intended to prevent rooms from heating up behind glazing, so that additional shading measures can be avoid where possible. This glass is admits a large proportion of light in the visible area of the spectrum, while transmitting little of the heat-generating solar radiation.

We distinguish between two modes of operation, which can also be combined:

_ <u>Absorption glass</u> is colored by adding metal oxides, so that parts of the incident radiant energy is absorbed and transformed into thermal energy. This is largely dispersed to the outside and in very small quantities inwards, with a delay. › Fig. 54
_ <u>Reflecting glass</u> works through metallic oxide coating that reflects large amounts of the incident radiant energy (UV and infrared radiation), but largely admits visible light. This usually selective coating is on the inside of the outer pane. › Fig. 55

Switched and variable glazing

More recent developments have addressed glazing that can be switched and varied. This type of glass is identified by its structure, the

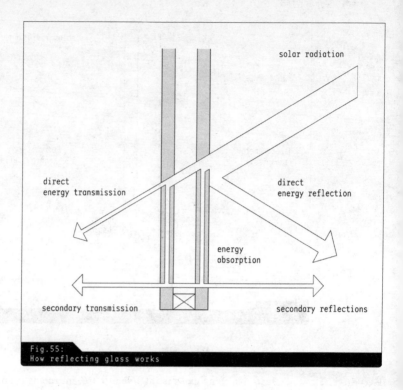

Fig.55:
How reflecting glass works

nature of the switchable layers and the way they are activated (electrochromic, gasochromic, photochromic and thermotropic systems). The degree of transmission can be altered using electric current, gas, thermal or solar radiation in parts of the glazing. In this way, the incident solar energy and the amount of incident light can be reduced or controlled according to the weather, or the time of day or year.

Soundproofing glass

The assessed sound reduction index (Rw) provides information about the sound insulation properties of glazing. The sound insulation value of an insulating glass unit can be improved by:

_ Increased pane weight
_ Different pane thicknesses (asymmetrical structure)
_ Using compound panes
_ Greater distance between panes with inert gas filling

Fire resistant glass

The glass used for fire resistance is always considered in the context of the frame and the way it is fastened to the shell. We distinguish between G- and F-category glazing.

G-category glazing prevents flames and combustion gases penetrating for the given time, but does not contain heat from the fire. G-glass can be supplied as special TSG or as laminated glass, in tempered soda-lime glass, for example.

F-category glazing contains flames and combustion gas penetration, and also the spread of heat radiation. This is always required in places where there is a fire escape route behind the fireproofed area. F-category

\\ Example:
The assessed sound reduction index uses a single value to define the sound reduction properties of a structural component or the sound insulation between room. This value for a structural element depends on frequency. We distinguish between two air sound reduction indices:

$R'w$: assessed sound reduction index in dB with sound transmission via flanking structural elements
Rw: assessed sound reduction index in dB without sound transmission via flanking structural elements
Characteristic value for air sound reduction
_ Walls, ceilings $R'w$
_ Doors, windows Rw

\\ Important:
Sound insulation value for windows
In Germany sound insulation for windows is divided into sound insulation classes.

Sound insulation class	Window	Assessed sound reduction index
Sound insulation class 1/2	Single window with insulation glass (4 / 12 air / 4)	25–34 dB
Sound insulation class 3	Single window with insulation glass (8 / 12 air / 4)	35–39 dB
Sound insulation class 4	Single window with insulation glass and cast resin filling (9 / 16 resin / 6)	40–44 dB
Sound insulation class 5	Single window with insulation glass and cast resin filling (13 / 16 resin / 6)	45–49 dB
	Double-glazed window, insulation glass (9 / 16 air / 4), single glass (6 mm)	45–49 dB
Sound insulation class 6	Box window: insulation glass (6 / 16 air / 4), single glass (6 mm)	› 50 dB

\\ Example:
DIN 4102 fire behavior of building materials and elements – fire retardant glazing

Structural elements	Fire resistance duration (minutes)
Walls, ceilings, beams, columns	F30–F180
Stairs, windows/glazing systems	F30, F60, F90, F120
	G30, G60, G90, G120
Fireproof partitions (doors, gates, flaps)	T30–T180

glazing has a gel in the gap between the panes. If fire breaks out, the g foams to create a tough, solid mass.

Fixed glazing can achieve fire resistance classes up to F90. As a ru an F-glass unit needs general approval in relation to building regulatior which also stipulate the number, type, and position of the attachme points to the shell.

Burglar protection glass

Burglar-resistant glazing is categorized in resistance classes (0– with special categories for bank and post office counters. They can combined with alarms. Safety glass is burglar- or even bullet-retardar Burglarproof glass is tested with a mechanically operated axe. Resistan classes are allocated by the number of blows needed to create a squa opening with edges of 400 mm.

FASTENING SYSTEMS

Glass mounting

Glazing beads are used to fix panes of glass in frames. They shou be fitted on the inside of a window to protect against break-ins, and th should be removable so that the pane can be replaced if the glass is broke Glazing beads ensure the integrity of glass and seal, absorb horizont loads (e.g. wind loads) and transmit these into the loadbearing frame cro sections.

The glazing bead must produce even pressure on the glass pane prevent the glass from breaking. Glazing beads can be screwed or clipp to the frame according to the system used.

Bars

Fixed glazing or opening windows can be divided by bars. The originate in the development history of glass. It used to be possible produce only small panes, and these were joined together with structur bars to produced large glazed areas. Now that glass is manufactured by the float process, bars are not required, but monument preservation, building regulations or special features of the place concerned can make the use of glazing with bars essential.

Placing glass – blocking

Panes have to be secured with glazing blocks to disperse the weight of the glass on the frame. Here we distinguish between loadbearing blocks, which support the glass in the frame, and spacer blocks, which secure the spacing between the edge of the glass and the fixed frame. In fixed glazing, the load is dispersed into the fixed frame and its anchorage in the shell. For opening windows, the load is dispersed via the opening frame and the points at which it is suspended (hinges, rollers, etc.). If the window is to function properly, it is essential that the frame and the opening leaf do not jam, twist or otherwise distort. The pane must not touch the frame at any point and the space between the rebate base and the gap between the panes must remain evenly distributed.

Vapor pressure must be equalized between the rebate space and the outside air, and it must be possible to remove condensation water. Vapor pressure equalization apertures must not be subject to direct wind pressure, so covering flaps may be necessary to protect the outlet openings. Vapor pressure equalization with the interior should be avoided, as this can lead to an accumulation of condensed water in the rebate space.

FITTINGS

Fittings are all the mechanical parts of the window that control opening and closing and secure assembly, fixing and use. › Fig. 56 Window and French window fittings combine the opening leaf and the fixed frame. The opening mechanism must be burglarproof, serve as a childproof lock, and ensure that wind and rain do not penetrate.

Fitting systems are supplied as component sets. They can be fitted concealed, (i.e. invisible), semi-concealed, or open, as in decorative fittings, for example. In the case of turn-and-tilt windows, for example, the moving connection of the window leaf with the fixed frame is created by the hinges, mainly drilled hinges, and by additional fittings:

- Stays: part of the fittings, moved and fixed when opening and closing the window. The leaf is open or shut by pushing along the locking points.
- Corner pivot rest: the window's rotation point. Supports the weight of the leaf.

Fig.56:
Window fitting

- <u>Scissor</u>: fixed at the top of the fixed frame and attached to the leaf fittings. With the pivot rest, the scissor forms the window's rotation axis, and controls switching from turning to tilting.
- <u>Locking plates</u>: the fittings on the fixed frame attached to the stay locking system. Windows are locked using all these plates, which also secures joint seals even in severe weather conditions.

Each opening type needs its own fittings.

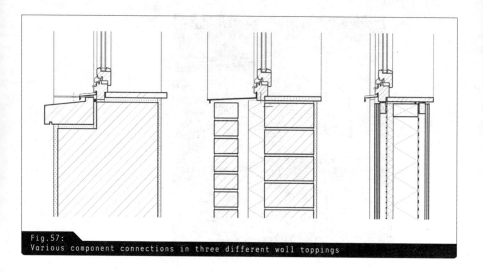

Fig.57:
Various component connections in three different wall toppings

FITTING STRUCTURAL ELEMENTS TOGETHER

BOTTOM OF THE WINDOW

The horizontal conclusion outside a window breast is usually formed by the sill. The following points must be borne in mind to ensure durability and freedom from damage:

- _ The window profile must protrude over the sill and have a drip cap (overlap) to keep it watertight even in driving rain.
- _ The sill must protrude at least 20 mm over the front outer wall edge and also have a drip cap (overlap) to prevent the soiling of the facade below.
- _ The sill must slope slightly outwards (at least 5° or 8%), so that water can run off.
- _ The sill must have an upward lip (e.g. sheet metal or slip-on profile) to prevent the structure from becoming damp. It makes sense to have a lateral overlap of the wall cladding to protect the joint between reveal and sill.
- _ Care should be taken with temperature-related expansion of the sill (length tolerance).

External windowsills

External windowsills can be made of sheet zinc or copper, glass, natural or artificial stone, an upright course of frost-resistant natural or artificial stone, clinker tiles, split tiles, or prefabricated aluminum profiles.

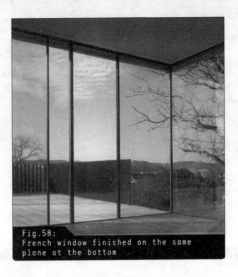

Fig. 58:
French window finished on the same plane at the bottom

Internal windowsills

Covering for internal window breasts or for radiator niches are usually realized in natural or artificial stone tiles in a bed of mortar or timber or derived timber products (on brackets where appropriate). > Fig. 57

French window

French windows differ from windows with breasts in the form of the lower part of the frame. Where appropriate, the bituminous damp- and waterproofing, terrace of balcony floor covering, and the internal floor fittings must be attached here. In the opening leaf of a French window, the lower horizontal profile is usually higher than for a window, to protect the glazing from splash water. As with a solid door, the bottom of a French window can also be realized on the same plane, by providing a gutter, for example. > Fig. 58

SEALS

Sealing prevents water from penetrating the building through the window structure, and reduces heat loss from uncontrolled air change. We distinguish between seals around the joints between the fixed frame and the shell (compression seals), seals between the fixed frame and the opening leaves (rebate seals), and seals between the frame and the glazing (joint seals).

Compression seals

The seals between the building and the fixed frame balance out tolerances and seal the joints on the outside against wind, driving rain and

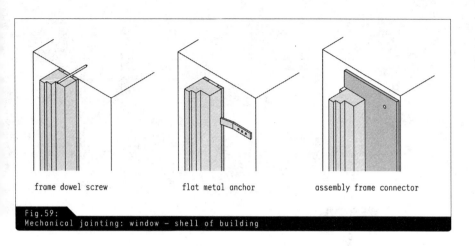

Fig.59:
Mechanical jointing: window – shell of building

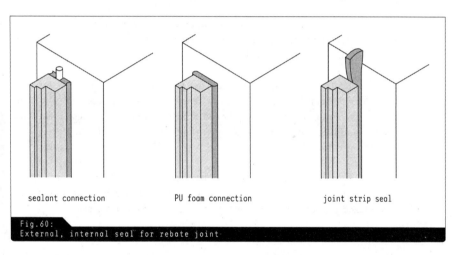

Fig.60:
External, internal seal for rebate joint

noise, and inside against water vapor diffusion. The window is fixed to the building mechanically in the compression seal area. The fixing elements transfer the vertical and horizontal forces generated. They must allow changes of shape in building and frame without damage. > **Figs 59 and 60**

Rebate seals

Seals in the leaf rebate between the fixed frame and the opening leaf frame are continuous and ensure adequate sound and heat insulation, and protection from moisture and drafts. Rebate seals consist of seal profiles (e.g. synthetic rubber or neoprene), which are glued into the window leaf

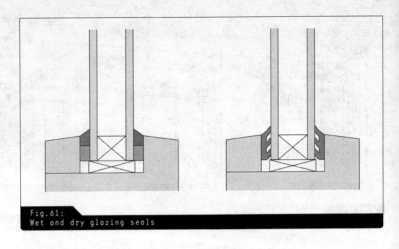

Fig.61:
Wet and dry glazing seals

rebate and where applicable the fixed frame, or pushed into custom-cut grooves. They are used in the form of single or even double and triple seals, and should ideally be replaceable.

Joint seals

Joint seals between frame and glazing unit are supplied for both sides, in continuous linear form. They are manufactured as wet seals in the form of glazing tape (made of sealants such as silicone, acrylate, polysulfide and polyurethane) or with prefabricated seal profiles (e.g. synthetic rubber). > Fig. 61

Prefabricated seal profiles are made by extrusion, so complex profiles can be supplied. Dry seals, unlike wet seals, can be built in with the glazed unit and the glass retention strip in a single process.

\\Note:
Connecting the window frame to the shell: Connection tolerances between window frame and reveal are absorbed by strip iron as an elastic element, dowel as a sliding connection, or U-shaped frame connectors as an adjustable link. Usually facade or window manufacturers provide a precise specification for the shell apertures, so that adequate tolerance can be allowed when the windows are being planned and made.

\\Example:
Precompressed joint strip (compression strip): Precompressed, impregnated foam sealing strip on a polyurethane base is supplied compressed to about 15% of the cross section and expands slowly after being fitted into the joint. It fits snugly into the edges of the joint, so that normal material movements at different temperatures are readily accommodated.

JOINTS

Vertical forces are usually dispersed by placing the windows on blocks or wedges that can also be deployed horizontally. Horizontal forces are dispersed by frame plugs, spacer screws, or flat metal anchors, called clamp irons, in the reveal area. The fixing devices must permit uninterrupted sealing by the compression joint.

Joints between fixed frame, masonry rebate and sill, and between sill and breast, are filled with mineral wool or PU foam to form a closed insulation level, but they should also be closed inside and out with weatherstrips. The standard PU foams cannot create durable wind- and vaporproof joints on their own. The best degree of seal – according to the facade cladding – can be achieved by continuously sealing the window frame outside with butyl strips.

Uneven and porous masonry should be rendered to prepare contacts.

\\ Important:
Window solutions:

Chicane
Prevents driving rain and moisture from penetrating. Fitted in the profile: by single or double rebating of frame and fixed frame; in the shell: precipitation is dispersed via the outer or inner rebate by the principle of overlapping several structural elements: reveal/rebate overlap the frame, the frame overlaps the opening leaf frame, the weather bar the sill, and the sill the wall cross section.

Tolerance, building movement
Between frame and shell: adjustable, elastic fixings and continuous joint with mastic joint sealant; between frame and fixed frame: using rebated joints.

Insulation
Positioning on the insulated plane of the wall, use of insulating glass, filling compression joints with mineral wool or PU foam.

Thermal separation
Occurs in profile and glass. In wood through the full cross section, in metal through additional-leaf construction or connection using plastic strips; in plastic using metal chamber profiles; for insulating glass units using spacers and air/gas fillings in the space between the panes.

Sealing
Between frame and glass: with sealing tape and seals of seal profiles; between frame and shell: by rebate, continuous foil sealing or joint sealing strips (see Figs 62 and 63).

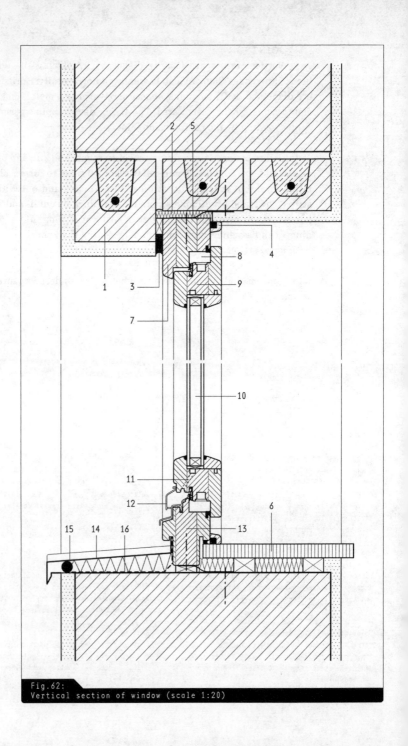

Fig.62:
Vertical section of window (scale 1:20)

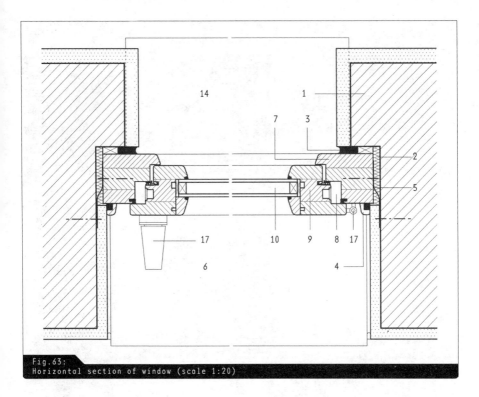

Fig.63:
Horizontal section of window (scale 1:20)

1 Shell connection — internal rebate
   Window lintel: here L-shell
   Window reveal: the external masonry shell forms a three-sided rebate (side-top)
2 Rebate insulation
   Mineral wool stuffing or plastic foam (PU foam)
3 Rebate seal
   Internal vapor seal: mastic sealant (e.g. silicon) on toroid (PU foam); outside: Wind seal with sealant tape (compression tape)
4 Covering strip
   Protection and finish for compression joint
5 Covering strip.
   Protection and finish for compression joint
6 Window lining/windowsill, slid on and screwed
7 Fixed frame. Profile: IV 78/78
8 Rebate space
   To accommodate fitting and closing elements
9 Opening leaf frame
   Profile: IV 78/78 with integral rebate and frame seals. The opening frame is rebated and surrounds the pane of glass on all sides.
10 Insulation glass unit
   8/16 mm, gap between panes 4 mm
11 Vapor pressure compensation
   Ventilation and drainage (condensation) of the rebate space via drilled holes
12 Rain protection bar
   Moisture and condensation water can run off through openings at the bottom of the bar
13 Bottom fixed frame profile
   Usually has a "drip cap". This overlaps the connection to the outer windowsill and protects the joint against precipitation.
14 Weather bar/external sill
   With at least 5° slope: weather bar: aluminum, squared off.
15 Lower rebate seal
   Mastic sealant (e.g. silicone) on glazing tape secures the windproof edge of the joint between the outer artificial stone windowsill and the assembly/fixed frame.
16 Windproof connection using windproof film.
   Film all round for windows without rebate.
17 Fitting

79

IN CONCLUSION

Apertures are a particularly topical architectural subject. Windows are the key to energy-saving planning and resource-friendly building for auxiliary solar heating and for lighting and ventilating rooms naturally.

Refurbishing old buildings is an important area of an architect's work. Here, apertures are important in terms of energetics, their design in particular. The huge building volumes of the post-war decades are hoping for a second chance, and frequently this ends in success or failure simply because of the way the windows are handled. Apertures in listed buildings and ensembles are clearly fewer in quantity, but all the more demanding in terms of the quality challenge they pose.

But design is not limited to successful dimensions and appropriate proportions. A window is a multifunctional structural element that can take up the familiar building repertoire again and also reinterpret it functionally. For example, Bruno Taut was able to add a lucid variant to the concept of the (kitchen) window for his "Onkel-Toms-Hütte" housing estate in Berlin (1926–31), based on his awareness of many models in regional building. Two opening leaves in each case, one an upright rectangle and the other in a somewhat squatter format, are placed above one another, thus creating a single window element in mirror image. This is not just an attractive design, but also enables continuous ventilation through the clearly staggered smaller leaves.

New developments in window mechanics, finish and style are always linked with design-related changes. These are contemporary features that give windows their particular character. The architectural field of apertures as a task for design and construction always requires awareness of structural and design contexts like these, and is of considerable significance when dealing with both old and new buildings.

APPENDIX

LITERATURE

Francis D.K. Ching: *Building Construction illustrated*, 3rd edition, John Wiley & Sons, 2004

Andrea Deplazes: *Constructing Architecture*, Birkhäuser Publishers, Basel 2005

Martin Evans: *Housing, climate and comfort*, Architectural Press, London 1980

Gerhard Hausladen, Petra Liedl, Michael de Saldanha, Christina Sager: *Climate Design*, Birkhäuser Publishers, Basel 2005

Thomas Herzog, Roland Krippner, Werner Lang: *Facade Construction Manual*, Birkhäuser Publishers, Basel 2004

Ernst Neufert, Peter Neufert: *Architects' Data*, 3rd edition, Blackwell Science, UK USA Australia 2004

Christian Schittich, Gerald Staib, Dieter Balkow, Matthias Schuler, Werner Sobek: *Glass Construction Manual*, Birkhäuser Publishers, Basel 2007

Andrew Watts: Modern Constuction: *Facades*, Springer, Vienna, New York 2004

PICTURE CREDITS

Figure page 8 (Th. Herzog)	Peter Bonfig
Figures 1, 2, 3, 4 (Ingenhoven Overdiek), 5 (F. O. Gehry), 6 (K. Melnikov), 7, 8, 13, 14, 18 (Ch. Mäckler), 20, 22, 23 (A. Reichel), 25 (P. C. von Seidlein), 30 (Kada Wittfeld), 41 (Kollhoff Timmermann), 42, 43, 45 (W. Gropius), 56 (Brinkman & van der Vlugt), page 78 (Wandel, Hoefer Lorch + Hirsch)	Roland Krippner
Figure page 12 (Lorenz & Musso), 24 (Lorenz & Musso), 33 (Lorenz & Musso), 52 (Lorenz & Musso)	Lorenz & Musso Architekten
Figure page 36 (Herzog & de Meuron), Figures 9, 10, 12, 31, 32, 33, 34, 35, 36, 37, 38, 39, 40, 44, 47, 48, 51, 53, 54, 55, 57, 61, 62, 63, Tables 1, 2, 3	Department of Building and Building Material Studies, TU Munich
Figures 11, 16, 17, 21, 26, 27, 28, 29	Herzog et al., Facade Construction Manual, 2004
Figures 15 (Sieveke+Weber), 19 (Sieveke+Weber)	Sieveke+Weber Architekten
Figure 46	Sonja Weber
Figure 49 left	Meister Fenster IV 72/75, Unilux AG, D-54528 Salmtal
Figure 49 right	Aluvogt, Bug-Alutechnik GmbH, D-88267 Vogt
Figure 49 centre	Hueck/Hartmann Aluminium Fensterserie Lamda 77XL, Eduard Hueck GmbH & Co. KG, D-58511 Lüdenscheid
Figure 50 left	Janisol primo, Jansen AG, CH-9463 Oberriet
Figure 50 right	System Quadro, Rekord § Fenster + Türen, D-25578 Dägeling
Figure 58	Fa. R&G Metallbau AG, CH-8548 Ellikon a. d. Thur
Figures 59, 60	Krüger, Konstruktiver Wärmeschutz, 2000
Work on drawings	Peter Sommersgutter

THE AUTHORS

Roland Krippner, Dr.-Ing. Architekt, academic assistant in the Industrial Design Department, TU Munich

Florian Musso, Univ. Prof. Dipl.-Ing., full professor in the Department of Building and Building Material Studies, TU Munich

Academic and editorial assistance: Dipl.-Ing. Sonja Weber and Dipl.-Ing. Thomas Lenzen, Department of Building and Building Material Studies, TU Munich

前言

开洞一般可以定义为"开敞的区域、洞口以及槽隙"等。建筑中的开洞则指的是墙体中预留的空洞位置。建筑的首要功能是形成一个受保护的空间,而建筑的第二项重要措施就是立面开洞,特别是对于具有使用功能的空间。

此外,开洞也是一项重要的建筑设计元素。由于开洞尺寸和比例的不同,与防水层、建筑面层的关系以及布置的不同,使得建筑中的不同开洞存在极大的不同。用来覆盖开洞的构件可以是活动的(窗户,门;二者也是建筑中的基本功能模块),也可以是固定的(镶嵌玻璃)。

在开洞设计中,窗户扮演的是最为中心的角色。建筑立面的窗户类型(构件、形式)、排列和分布相当于建筑的名片。窗户的分格和镶嵌玻璃的内部结构则是影响窗户整体效果的另一个重要因素。从窗户的形式上也能够清楚地辨别出其风格上的改进和加工工艺的水平。

除窗户和门本身之外,其控制系统也是开洞系统的附加设备。控制系统能够精确地控制建筑的采光、通风以及热量传递。根据天气条件的不同,用户可以利用控制系统对室内空间进行调节,以达到自己所需要的舒适程度。

在窗户的设计中,存在多种不同的体系类型和构件形式可供选用。在进行开洞设计的时候,为了根据特定的气候条件找到有效的设计方案,就必须对它们的基本功能和构造条件有一定的了解。此时,窗户的设计特点与建筑的功能需求和结构质量密切相关。

立面开洞还需要具备与立面相同的功能(抵御寒冷、潮湿、噪音、火灾以及外部入侵等),但是此时开洞位置成为了建筑围护结构中的保温薄弱点,所以从建筑节能的角度需要对开洞处进行特殊处理。开洞处有时可以起到出入口的功能,同时还可以划分墙体以及室内空间中用于采光和通风的区域,以及确保室内的人能够往室外眺望。

因此,开洞处的通透性应该是可以调节的,可以通过特殊的封闭构件对开洞处的热量、光线以及空气流通情况根据建筑所处的地理位置(不同的气候条件)以及住户所需的舒适度(稳定的室内气候)进行调节。而调节的效果则与太阳位置、室内空间的考虑以及使用情况有关。

图1:
天然石材墙体中的开洞

图2:
实体木墙（圆木结构）中的开洞

图3:
圆形窗户（位于20世纪50年代的一个阁楼中）

图4:
通风口

图5:
全层高开洞

图6:
蜂窝状开洞

从建筑历史的角度出发，门比窗户的出现时间更早，它为建筑物提供出入口。广义的门还包括大门，通常大门的尺寸较大，而且往往用于机动车通行。在大多数人看来，建筑立面开洞往往指的就是窗户，主要是采用一些透明的或者半透明的材料，用于采光（即使在窗户关闭的时候）和通风（当窗户开敞的时候）。

　　封闭开洞的最简单方式是采用固定镶嵌玻璃。此时，可以把窗户的采光和通风功能分别实现（比如，采用镶嵌玻璃结合通风板的形式）。

　　立面开洞的公共立面（朝向建筑外侧）往往需要满足其影响力的要求。开洞的数量、尺寸通常能够反映出建筑拥有者和居住者的社会地位。而开洞的内侧立面则具有一定私密性和亲切性。门窗的布置同样也能够影响建筑的室内效果。如果房间的窗户布置得当，可以使房间显得比实际尺寸更加宽敞，同时通风效果更好。

开洞的功能

保护功能

多个世纪以来，对于窗户的基本功能要求没有发生变化，但在保温、隔音、防火以及构件与墙体之间的气密性方面的要求却有了较大的改变。对于窗户性质以及效率方面的要求主要取决于：建筑所处的位置——地形、建筑的朝向、建筑的高度以及受主要风向的影响程度。

保温和隔音

窗户的保温和隔音效果主要取决于窗框的材料选用、窗框的厚度或者结构形式以及玻璃的类型、厚度以及安装方式等。

窗户必须满足保护室内抵抗寒冷的最低保温要求。即使采用较薄的窗框和玻璃，窗户也必须缓冲窗内外的温差峰值。为了避免热桥的产生，窗户与墙体隔热层之间的搭接长度应该尽可能长；同时，为了避免产生冷凝和发霉现象，应确保窗内侧的温度相对较高。（详见"窗户的组成"一章中"玻璃"部分内容）

外墙应该起到相邻空间和室外之间的隔音作用。大多数情况下，室内空间应该避免外部噪音的影响（比如飞机、道路交通灯，等等），反过来也一样。

保温和隔音功能可以通过采用特殊的玻璃（保温、隔音玻璃）或者不同类型的窗户（冬季窗、双层玻璃窗户、盒型窗）来实现。改善隔热性能，能够提高窗户玻璃内表面的温度，从而降低居住者对窗户附近的冷空气的不舒适感。要想进一步提高舒适度要求，就需要在窗户构件中采用加热技术。另外，采用组合窗框、绝热木质窗框以及隔热金属构件都能够有效提高窗户的整体效果。

提示：

外墙和外墙窗户玻璃，特别是玻璃的内表面温度，将影响房屋保温和舒适度。居住者所感受到的是空气和表面的平均气温。通过墙体材料和结构的选择，以及（或者）窗户形式的改进，可以确保室内温度在居住者感知的适宜温度范围之内。

提示：

冷凝的形成

当（室内的）空气中的水蒸气冷却到了液态，就形成了冷凝。所以说，室内表面，尤其是玻璃以及搭接部位，应尽可能保持较高的温度。

防潮和气密性	除了冬季保温之外，还要注意防止夏季的炎热或者温度过高的情况。这方面的考虑主要取决于建筑立面的朝向以及开洞的尺寸大小。 窗户应该避免受到雨淋（水溅）以及潮湿，否则会损坏建筑物的面层。 为了避免漏风和控制空气的流通，窗户必须设有密闭设施和紧固件。这些构件主要设置在窗户与建筑的连接处，同时也设置在窗框和窗扇之间。它们可以改进窗户的绝热效果，能够减少通风过程中的热量损失，起到更好的节能效果。 室内的湿气将对窗户以及连接部位造成影响，而这种影响的程度一般是与天气相关的。考虑到水蒸气的扩散，对功能层的设计应该服从内侧厚于外层的原则，以确保湿气能够从建筑的内部向外部扩散。
遮挡与眩光保护	大尺寸的开洞需要进行视线遮挡以确保内部空间的私密性。 过大的光密度反差将会引起眩光现象。特别是对于监测性质的工作，眩光将造成非常大的干扰。防眩光系统则可以对外来光辐射进行控制，从而降低视野与电脑屏幕之间的亮度差。 防眩光装置安装在窗户内侧，但不应该完全遮挡日光或者将视线完全阻断。最理想的做法是，防眩光装置能够从开洞处的下端进行提升并固定在任何理想的高度。见图 7 遮光装置则可以放置在窗户内侧、外侧以及房间内部，或者是采用百叶窗的形式。
防火	建筑内部不允许发生火灾，如果一旦着火，最重要的是防止火势扩散。火势可以通过开洞在房屋和楼层之间进行扩散，所以窗框和窗扇需要满足防火的要求以防止火势扩散。可以通过窗框和窗扇玻璃材料的选择来满足以上要求。

> **重要：**
> 从人体工程学的角度，在建筑物的窗户和天窗建造中进行防眩光和防晒设计是一项非常重要的工作，从而避免房间直间暴露在阳光的照射中。可参考《EU Directive 90/270/EEC》、《ISO 9241》，以及各国国家规范中的相关规定和建议。

图7:
半透明防眩光系统

图8:
防止坠落

防冲击和 防坠落	对于高层窗户玻璃，尤其是商店橱窗、全景玻璃窗，在窗户设计时需要考虑冲击荷载的影响，必要的时候还应采用设有固位结构的特殊玻璃。 　　窗栏的高度（取决于允许的最大跌落高度）必须符合建筑规范的相关要求，以防止从开窗处或者大型镶嵌玻璃以及全景玻璃门处发生坠落。(详见"洞口的组成"一章中"构件"部分内容)可以通过采用窗栏、固定玻璃窗或者横档的方式来实现。)见图8
防盗	外墙中容易进入室内的位置（比如，与阳台相邻的首层地面以及窗户位置）一般需要采用特殊的保护措施以达到防盗的效果。由于普通的窗户是不具备防盗功能的，所以普通窗户和玻璃门很容易被撬开。为了防盗，需要安装与窗户构件相匹配的整套防盗构件（窗框与洞口构件的一体化、防盗配件、防盗玻璃等），或者通过安装安全装置、窗格栅以及开启装置等。
P18	**控制功能** 　　在进行开洞设计时——开洞的大小、布置等——需要满足一系列的要求，有时候会出现相互冲突的情况，但总体上需要达到协调一致的效果。 　　开洞的大小不仅影响视野的范围和与外部空间的接触，更重要的是将影响照明功能和对太阳能的直接利用。
开洞面积	在决定"开洞面积"（开洞尺寸以及窗框所占比例）的时候，最

重要的是开洞朝向的方位以及其绝对尺寸。

增加开洞尺寸意味着：

——接受更多的日光

——接受更多的辐射

——夏季的炎热问题

——保温效果降低

——清洁需求增强

墙面和开洞几何形状的设计与其内部的房屋设计紧密相关。二者都会对采光量、通风以及用户对外部空间的视野造成影响。

开洞的位置选择与开洞的水平和竖向功能均有关系，有时家具的不同选择可以改变开洞之间的水平关系。

在竖直方向，外墙分为以下几个区域：

——气窗（横楣上方的顶部窗口）

——窗口（房间内部的视野范围）

——窗栏

视线　　通过采用透明、无畸变以及理想的中性色彩建筑材料（比如玻璃或者塑料等），能够让建筑内部与外部取得视线联系。通过窗户的尺寸和布置可以开阔用户的视野——比如采用条形状、全景窗，以及落地窗等；同时也可以对视线进行强烈约束和限制——比如采用小尺寸的开洞以及精确地固定开洞的位置。〉见图9

视平线　　在确定开洞位置的时候应该考虑人坐着或站立时的视平线高度。人体比例如身材尺寸、视野和活动姿态（站立、坐着、躺着）是影响视平线高度的关键因素。以下视平线或者视线高度平均值可供不同情况参考：

——站立大约 150-170cm

提示：

大多数国家对相邻建筑物之间的立面开洞进行了相关规定。比如，建筑物之间的相邻距离满足一定条件方可开洞，一般来说这个距离不得小于5m。

重要：

以下尺寸可作为在客厅内视线的划分依据（地面以上高度）：

——窗户顶部边缘位置 ≥ 2.2m

——窗户底部边缘位置 ≥ 0.95m

——窗户的宽度 ≥ 房间宽度的 55%

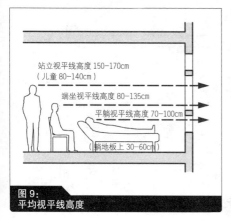

图9：
平均视平线高度

图10：
吧台区域的视线

— 坐着（工作时）大约 80-135cm

— 平躺大约 70-100cm

开洞面积应该根据用户在室内的活动和姿态进行细分，所布置的水平或者竖向建筑构件不要对视平线与外界的联系造成阻碍。见图10

通常人们在渴望与外部空间取得视觉联系的同时，也希望能够获得外界的新鲜空间。打个比方，我们更希望站立在一个可以打开的窗户前。所以说，窗户的开启装置应该在用户方便打开的范围之内。

采光

从生理学角度和节能角度来说，自然采光都非常重要。通过光线的强弱，室内阴影的位置以及光线的颜色可以感知到太阳光每天以及季节性的变化。充足的自然光对身体健康以及良好的工作效率非常重要。

对于客厅而言，相关建筑规范规定其开洞的最小面积是室内地面（可用空间）面积的八分之一。同时，该最小值还与房间的进深有关，如果只能从房间的一边进行采光，那么采光面积的比例将随着房屋进深的增大而减小。

外部影响因素

外部影响因素包括：

— 定位与朝向

— 日间与季节性变化

— 光线强度和光的颜色

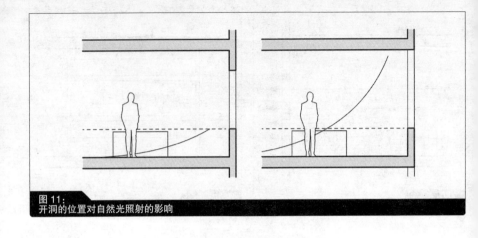

图 11：
开洞的位置对自然光照射的影响

— 附近物体的阴影（植被、其他建筑物等）

对于房间采光，开洞的上边缘位置尤其重要。窗户位置越高，进入房间的光线越多。在开洞面积相同的情况下，采光系数会随着开洞上边缘高度的提高而增大。另外，窗下墙位置处（指 0.85m 高度以下位置）的开洞则对增加采光的作用有限。〉见图 11

对于房间采光而言，外墙中的开洞位置比开洞的相对尺寸更加重要。一般情况下，对于相同尺寸的外墙开洞，通过垂直外墙中开洞的光线将比通过平屋顶天窗开洞的光线减少 5 倍以上。而光线的亮度则取决于室内表面对光线的反射程度，此时的主要影响因素则是房间的主导颜色。

工作区域应该尽量靠近窗户布置，此时需要注意的是光线进入的方向（避免用户还需要采用其他的照明方式）和用户的视线方向（通常是与洞口的朝向平行）。

重要：
采光系数（D）指的是在正常情况下的入射光百分比，其表示的是室内与室外（只有散射光的情况下）的照明强度比值。

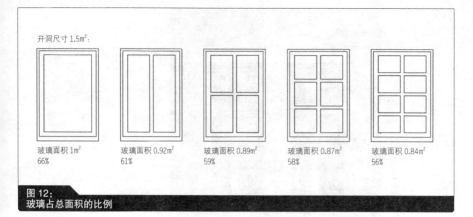

图12:
玻璃占总面积的比例

当光线进入房间的时候，部分光线会被结构构件以及尘土所遮挡。同时，高达40%的窗户面积将被窗框、立柱、横梁和栅栏所占据。〉见图12和图13

结构横向构件，比如保护墙以及开洞上方的突出构件（窗罩、屋顶悬挑以及阳台等）将影响房屋的采光并改变光线在室内的分布。因此，这些不透光的结构构件，应该能够尽可能地反射阳光，比如将他们漆成浅色以便反光。另外，通过在开洞前方放置光导系统和使用高反射率的材料作为天花板能够更好地将自然光带入到房间深处。

太阳能的利用

对于休闲性质的房屋，太阳光则非常重要。另外，由于"温室效应"的作用，通过开洞处玻璃的光线可以作为房屋采暖系统的补充。

将开洞及其后方空间作为一个整体，则可以看做是一个简单的太阳能收集和存储系统。以下四个重要因素将决定能够直接利用的太阳能：气候和地理因素、开洞的朝向、开洞的倾斜角度以及遮挡情况。

在欧洲中部地区，太阳能已经开始使用，但不论是从每天或者季节性角度出发，都无法满足建筑的热量需求。对于一个保温条件良好的房屋，高达三分之一的热量需求可以通过南向表面的阳光直射来获取。如果开洞方向是朝东或者朝西（夏季时阳光照射强度特别高），应该确保开洞的尺寸比例不要超过40%，以确保房屋的热平衡。

图 13:
新艺术风格的窗户

> 💡 辐射量与开洞面积成正比,若玻璃面积过大可能会引发温度过高的问题。

采光面积、开洞尺寸、热量需求以及保温材料之间应该处于一个相互平衡的状态。即使是在采暖期间,开洞处 50% 以上面积可能都是不需要的。此时,可以通过设置遮阳措施(比如,凸出的结构构件),即使无冷却系统也能确保室内处于舒适的温度范围内。

> 📎 太阳能的收集还会受到窗户结构以及与用户相关因素的影响,所以也存在窄窗和网格状类型的窗户。另外,窗网和窗帘也会减少太阳能的收集,而内部遮光和防眩光系统将进一步减少太阳能的收集。

事实上,通过建筑特征的修改可以将太阳辐射降低到最初理论值的一半,而通过用户相关因素的调整,可以进一步降低至最初值

> 💡 **重要:**
> 对于具有大面积玻璃窗的办公室,可能同时存在太阳能过热和内部热负荷(人工照明、办公设备以及人员活动)较大的情况,这意味着需要进行大量的冷却工作。

> 📎 **提示:**
> 太阳能获得量指的是通过窗户或者其他透明/半透明结构构件进入建筑内部的太阳能总量。太阳能获得量有助于提高建筑和室内空气的温度,从而减少建筑采暖的需求。但是,可以利用的太阳能只是获得量中的一部分,其他部分将消散在建筑的周边区域。

的三分之一。

通风

除照明之外，通风是开洞的另一个重要功能。通风对于居住者的舒适感和健康，以及建筑物的保护来说都是一个基本条件。通风指的是室内空气与室外新鲜空气之间的交换。传统上，这个过程是通过外墙中的窗户来完成的（理想的空气交换），或者通过窗户与开洞之间的连接部位以及防眩光装置或者百叶窗来完成的（非理想的空气交换）。

由空气的内外部压力差引起的空气交换，被称为自由通风或者自然通风（也称冲击通风）。内部空气的交换有助于为室内提供新鲜空气，同时减少水蒸气、气味以及产生的二氧化碳，并很大程度地提高舒适度水平。

窗户通风

空气交换的程度与窗户的面积和开启方式有关。其他因素，比如其他开启的窗户（例如，位于对面墙上）、室内门的开启程度、百叶窗滚轮的位置、板条的状态都可能增加室内的通风。根据用户习惯的差异，建议空气交换的频率在每小时0.5–1次。另外，室内空气的流通速度同样能够影响舒适度，其速度上限为0.2m/s。

可独立开启窗户的优势是用户可以自由地调节室内空气。可独立开启的窗户不应该触碰到其他结构构件（比如遮阳板、防眩系统、遮光板以及结构墙、柱等），同时也不应该影响内部空间的使用。如果只在房间的一侧设置窗户，则认为自然通风所到达的房间深度不超过开洞净高的2.5倍。

窗户并不一定需要承担房间的通风需求，特别对于固定三层隔热玻璃窗以及大幅玻璃窗，考虑到结构构件的重量，将固定玻璃构件与

提示：

换气频率n（1/h）指的是一小时内室内或者建筑物内空气交换的程度。

例如：n=10/h

指的是室内或建筑物内一小时空间交换的次数为10次。

重要：

空气交换（AC）

窗户位置	每小时空气交换次数
门窗关闭	0.1–0.3
窗户倾斜，百叶窗关闭	0.3–1.5
窗户倾斜	0.8–4
窗户半开	5–10
窗户全开	9–15
对面窗户或者中间门全开	可达到40

图 14：
设有不透明通风口的玻璃系统

绝热效果较好的轻质通风门分开设置可能是一个更好的选择。〉见图 14

在进行窗户通风的时候,用户的行为可能会引发一些问题。比如,人们经常忘记按时通风,导致室内空气流通不够充分。另外,由于开窗通风时空气的流动无法得到控制,可能会导致冬天建筑的热量损失增大以及夏天时需要进行额外的冷却。

通风造成的能量损失成为了建筑耗能中一项重要的预算,因此需要在外墙建筑中提高节能技术。通过可控的通风系统可以避免不合理的通风路径,但同时也导致了个体化的空气调节功能,尤其是对于安装了进风和排风系统的被动式建筑。

结构接缝处的空气流通

对于许多建筑(特别是年代较久的建筑),在结构接缝处存在额外的空气流通。接缝处的空气流通一般无法得到控制,有时甚至不

技巧:
小面积的开洞可以很好地进行调节,将其沿垂直方向尽可能间距较远地布置,特别适合进行(长期)通风,同时也可以防止过度的通风和热量损失。从这个角度出发,在英语国家常用的垂直推拉窗比欧洲地区的内开内倒窗户更有优势。

会对气流产生影响。但接缝处可能存在水汽的冷凝现象,而对结构构件产生损坏。不过对于新建建筑而言则不存在这个问题,通过改进接缝处交界面的密封性可以达到完全的气密性。

P26

边缘构件

开洞的周边设有结构构件,而它们的功能则与所在的位置有关。通常,出于结构材料性质和建造模数的考虑,边缘构件的首选形式为矩形。但矩形并不是必须的,通过改变变形构件的形式,可以改善采光、窗外风景以及对太阳能的直接利用。

特别是对于采用较厚墙体的建筑,开洞同时还是室内墙面与室外墙面以及防水层的分隔。所以,此时开洞处的边缘构件可能会根据窗户的位置做成不同的三维形状。

在现代建筑中,由于节能建筑和新的建筑技术的需求,意味着墙厚不再是朝着更薄的方向发展。相反地,由于更高的保温隔热要求以及隔热层厚度的增加,墙面再次变厚。所以,开洞边缘的处理方式再次成为了研究热点。

墙厚

对于相同的开洞面积,墙厚、外观设计、采光量以及太阳能的利用等因素之间相互影响。另外,建筑的不同类型和建造方法可以通过扩张或者收缩视觉链接的方式对从室内向室外的视线造成影响。〉见图 15 和图 16

对于非承载立面,上部墙体结构的厚度较薄,因此在太阳光照射时的阴影范围也较小。在考虑墙厚的时候,需要考虑洞口处立柱

图 15:
边缘的几何设计

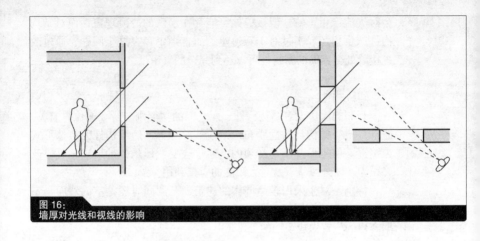

图16：
墙厚对光线和视线的影响

和滑轨超出墙体的部分，同时可能还需要考虑玻璃的厚度。

几何设计　　边缘的厚度主要取决于墙体的结构。边缘的几何设计应该与边缘构件（比如窗户或者固定玻璃等）的槽边统一考虑。

将洞口边缘做成坡口形式，可以在开洞面积较小的时候增加采光量，或者让开洞面积从视觉上显得更大一些。窗户可以做成向内或者向外倾斜的形式来调节采光量，另外还可以通过在洞口边缘涂刷反光能力强的浅色涂料来增加室内的采光。

开洞边缘的形式可以做成周边一致、各边对称、非对称以及各边各部相同的形式。不考虑外墙的材料质量，可以在清水墙体或者适当装修的混凝土墙体和天然石墙体中采用坡口形式的洞口边缘。

向外开坡口　　窗顶过梁处采用向外开坡口的形式可以增加顶部进入室内的采光量。窗台处采用向外放坡的形式，可以提高窗台密封处的排水性能，同时还能够增加室内外空间之间的联系——对于高层建筑尤其如此。同时，窗栏还可以起到遮挡视线的作用。〉见图17

向内开坡口　　向内开坡口的做法可以降低墙体与开洞处的光强对比，因此能够起到防眩光的作用。洞口的直边将增强光线的剪影效果。一般窗户内侧与室内墙面之间的剖面宽度较小。在天气晴朗或者房间暴露在阳光直射的情况下，这种开坡口方式能够确保室内过渡为一个中等光强的空间。根据开坡口宽度和角度的不同，能够确保没有眩光或者只有轻微的眩光发生。

"将窗框设计成为具有一定宽度、向内开坡口边缘的形式：大约一英尺宽、坡口与窗户平面夹角为50-60度之间，能够确保白天的

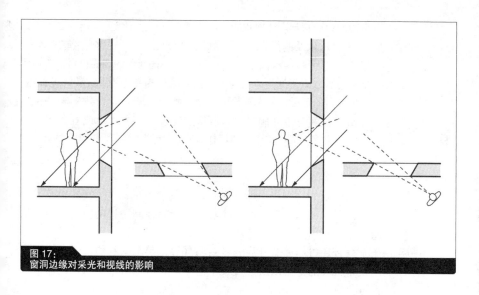

图 17:
窗洞边缘对采光和视线的影响

时候窗户与室内墙体之间的光强过渡比较柔和。"(参考 Christopher Alexander 等撰写的《*A Pattern Language*》一书,牛津大学出版社,纽约,1977 年,第 1055 页)

与传统做法不同,现代建筑中通常采用一些非对称的窗洞口做法。这些特殊的做法通常与当地的气候条件或者城市特征有关。)见图 18 和图 19

在进行洞口边缘几何形状的最后涂饰时,需要特别注意洞口的尺

图 18:
单边放坡窗洞

图 19:
两边放坡窗洞

寸以及洞口面积与墙体面积的比值是否正确。采用参差错落或者步进式的结构构件，以及偏置的洞口边缘做法能够使墙体外表面产生雕塑效果。如果增设光影设置或者采用不同性质以及不同颜色的墙面材料，能够更加增强这种效果。通过设置窗外突出的结构构件（侧壁或装饰板），能够增加窗台的宽度，但无法减小（详见"洞口的组成"一章"构件"部分内容）。

虽然目前对于开洞尺寸、洞口边缘设计、采光量或者太阳能的利用没有太多的量化指标，但对洞口的设计提出了大量相关要求。

P29

控制设备

洞口处的控制设备能够对室内空间所受的天气/气候影响进行人工或者自动的调节。这种控制设备能够控制洞口的空气流通，从而根据季节和每日的时间对空气质量、室内外温湿度差异进行调节；控制设备还可以完全开启或者封闭外界的热辐射和热传递。根据控制设备机理的不同以及强度要求的不同，可以对不同的控制设备进行组合使用。

随着开洞尺寸的增大，还需要寻找更加合适的遮挡材料以及控制结构形式，以确保实现理想的洞口通透程度。早先时候，通常采用动物毛皮、织物或者纸张来进行遮挡。后来，逐步被可以转动或者滑动的活动板条（木质）所取代，或者采用半透明的有机材料进行覆盖。这些简单的木质活动板条经过不断的改进，发展出多种改进形式或者新的形式。〉见图20

控制设备还能够采用机械自动控制。通过传感器的设置，可以实现根据天气情况而对控制设备的自动控制。此时，应该确保用户对控制设备进行个人控制的优先权。通过不同构件组合和优化，能够实现所需的舒适度和能量消耗。

位置设置

控制设备的位置设置能够对其功能造成直接的影响。相关因素

示例：
关于窗外突出构件在建筑中的运用可以在法兰克福建筑师克里斯托弗·马克勒（Christoph Mäckler）以及斯图加特的Lederer Ragnarsdóttir Oei建筑师工作室的作品中找到。

图20：
设有可活动条板的百叶窗

包括：

— 与洞口的相关位置（上部 / 中部 / 底部 / 一边或者多边）

— 在外墙的相关位置（位于墙体外侧远离洞口 / 位于墙体外侧 / 位于窗户平面 / 位于墙体内侧）

功能性质　　不同的控制设备在技术原理和材料选用上存在不同，在操作方式和活动程度上差异更加明显。另外，设备的活动方式和方向也对功能性质存在一定的影响。

相似的控制设备存在多种不同形式的装饰做法。总的来说，一共存在以下三种类型特征：

— 通透性

— 活动方式（通过相关构件）

— 包装尺寸

通透性　　影响控制设备品质要求的关键因素包括洞口所需的换气、采光以及太阳能利用方式和通透程度，可以将控制设备设置在半开半闭的位置，然后再设置所需的通透程度（比如从完全开到仅开一条小缝）。

活动方式　　可活动控制设备可以细分为可拆除的以及活动的两种，具体可以用发挥作用的时间来划分：包括固定安装但可以临时拆除的，比如附加双层玻璃、百叶窗等；或者永久安装的，比如可折叠百叶窗以及卷帘等。

在本册书中提到的可活动控制设备指的是永久安装的可活动控制设备。也有一些控制设备是完全不可活动的（比如可开关的特殊玻璃、气致变色玻璃以及电致变色玻璃等）(详见"窗户的组成"一章中"特殊玻璃"部分内容)。

控制设备可能由不同的构件组成，其活动方式各不相同，同时创造出的通透方式和通透程度也各不相同。

包装尺寸　　控制设备的包装尺寸，包括使用过程中尺寸的变化情况，对控制设备的选用具有重要的影响。在使用过程中，其尺寸可能是保持不变的（比如，铰式百叶窗），也可能是尺寸减小的（比如，可折叠百叶窗，或者更明显的窗户卷帘，其收起后的高度仅为其覆盖高度的6%~10%），尺寸还会对控制设备的操作方式产生直接影响。

活动性质与方式　　控制设备的活动方式通常是一系列运动原理的组合。不同的运动原理加上可能的运动方向，可以组合出多种不同可能的活动方式。如果能够对不同的光、声、热等条件做出有效的反应，就可能达到理想的舒适程度。控制设备可能的活动方式如下：

— 绕竖向轴运动（转动，卷帘等）
— 绕水平轴运动（转动，水平板条等）
— 构件尺寸不改变，构件位置完全发生变化（平移，滑动百叶窗）
— 构件尺寸和位置完全发生变化（变形，卷帘等）

活动方式还可以根据操作方式以及活动所需的相对空间来进行分类——通常是向外/向内以及向上/向下：

— 转动：向外开启/向内开启
— 折叠：向外开启/向内开启
— 滑动：水平向右/向左/竖向（向上、向下）〉见图21

制作形式　　用来封闭开洞的控制设备的最简单形式是百叶窗，百叶窗通常采用木质或者其他简单易得且工作性能较好的材料制作而成。另外，也有采用天然石材制作百叶窗的先例，而且自19世纪以来，很多地方也采用了金属百叶窗。最初，百叶窗是用来替代窗户的，而自15世纪以来，百叶窗逐渐转化成了窗户的附属构件。

> **示例：**
> 如今，百叶窗越来越体现出设备的精密性以及构造的复杂程度。从最初的不透明木质百叶窗，到后来的半透明以及透明材料，用来确保采光和向外的视野。大约在1700年前后，平板式百叶窗逐渐被斜向固定安装的条板所取代。再后来，斜向的条板可以人为地进行活动调整——采用一根木条或者金属棒能够对采光量进行调整。在早先时候，存在将百叶窗进行功能分区的做法，这样可以将采光、视野、通风以及遮挡功能分别进行调整。

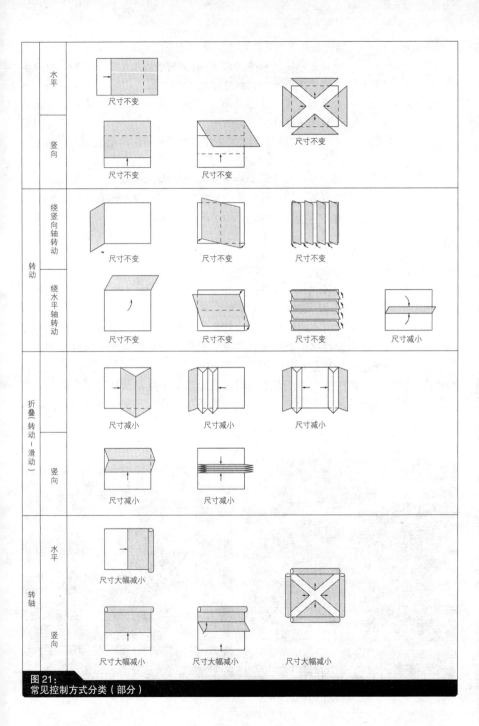

图 21：
常见控制方式分类（部分）

| 活动方式与 | 控制设备的类型可以根据活动方式与固定方式进行划分： |
| 固定方式 | |

— <u>推拉百叶窗</u>（水平滑动），通常用于较小尺寸的开洞，安放于洞口内侧或者外侧的轨道之上〉见图22

— <u>升降百叶窗</u>（竖向滑动），安放于洞口的顶部或者底部，通常设置在墙体结构内部

— <u>平开百叶窗</u>（转动），安放于洞口两侧，其变换形式包括竖向板条百叶窗〉见图20

— <u>箱式百叶窗</u>（折叠和倾斜），于洞口上部或者下部铰接

— <u>折叠百叶窗</u>（同时滑动和转动），洞口侧向铰接〉见图23与图24

— <u>滚轴百叶窗</u>（由两端挂在绳索或者铁链上的窄条组成，或者采用薄膜形式），其遮阳系统（通过滚轴转动）安放在洞口上部，并且部分嵌入至墙体结构中

— <u>条板或者百叶帘</u>（由两端与绳索固定连接的窄条组成，同时发生滑动和转动），安放在洞口的上部，且部分嵌入至墙体结构中〉见图25

几乎所有的常见建筑材料种类都可以制作控制设备。如果控制设备中含有多个运动方式不同的组件，则需要确保不同组件之间工作正常，不发生相互干扰。

| 构件尺寸 | 构件的大小或者截面尺寸主要取决于控制设备的尺寸和活动方式。因此对于推拉百叶窗或者水平折叠百叶窗，则适合采用竖向形 |

图22：
轻质金属水平滑动构件

图23：
装有木板条的折叠-转动百叶窗

108

图24：
木质遮光折叠-转动百叶窗

图25：
轻质金属百叶帘

式的构件来传递百叶窗配件和支承结构的荷载，而对于箱式百叶窗则适合采用水平形式的构件。如果推拉百叶窗的长宽比过大，则在推动时非常容易发生堵塞现象。另外，还需要注意条板以及滚轴遮阳系统等条形构件的宽度，以免发生构件弯曲。

应用实例：
遮阳设施

在进行控制设备的选择时，重要的前提是对控制设备工作原理以及控制设备与天气之间相互作用的了解。如果是为了遮挡日光（为开洞处提供阴影），必须要求控制设备能够对不同的天气条件以及太阳位置的日常变化、季节变化做出相应的反应。

以南向开洞为例，可以根据提供阴影的不同原理分为以下类型：
— 完全阴影，直接将开洞处覆盖：这种方式完全阻断了与外界的视线接触，所以室内需要提供人工照明；
— 采用半透明结构（开口或者网眼金属板等）：遮挡之后允许与外界进行有限的视线联系，同时进行一定程度的采光。）见图26

条板结构

更有效的做法是采用由一定数量小型构件组成的条板结构将开洞区域划分成若干区域。当太阳位置移动时，条板随之调整，并能够确保与外界的视线联系。其最好的方式是采用互不相连的独立控制系统。条板结构能够提供阴影，同时还能够在视线高度位置提供向外的视野。入射的太阳光将被窗户上半部处的板条遮挡，这种板条结构能够很好地遮挡阳光直射，同时还能够采集自然光和确保与外界的视线接触，通过采用半透明的条板，能够进一步优化这种结

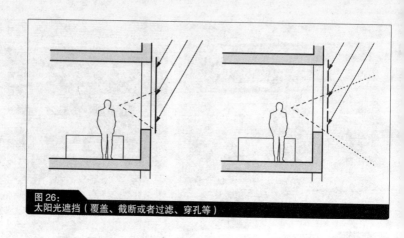

图 26:
太阳光遮挡（覆盖、截断或者过滤、穿孔等）

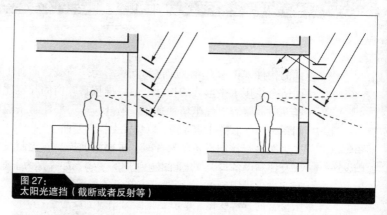

图 27:
太阳光遮挡（截断或者反射等）

构的性能。〉见图 27

方向定位　　条板结构主要有两种方向定位，主要取决于窗户的朝向以及太阳的相对位置：

— <u>水平条板</u>能够防止南向太阳高度角较高的阳光入射。当太阳高度角降低（到了东方或者西方）时，进行太阳光遮挡就更有意义；

— 如果开洞方向朝东或者朝西，则采用<u>竖向条板</u>更能够防止阳光透射。

在提供阴影的情况下，仍然能够确保向外的视野。〉见图 28～图 30

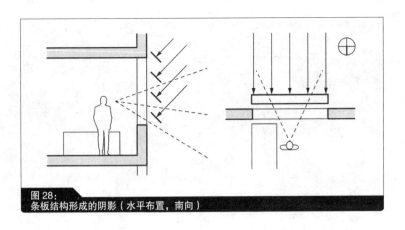

图28：
条板结构形成的阴影（水平布置，南向）

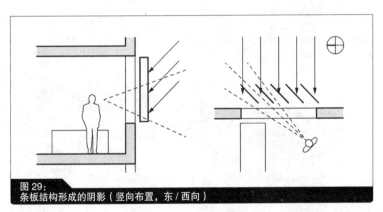

图29：
条板结构形成的阴影（竖向布置，东/西向）

图30：
竖向轻质金属板条

洞口的组成

构件

过梁　　过梁是位于墙体开洞最上端的构件。过梁跨越洞口两端，将洞口上部的荷载传递到两侧的承重墙上。过梁最简单的形式是采用一根承受弯矩的梁（比如，木质梁、钢筋混凝土梁、钢梁等）。此时，洞口尺寸主要受到过梁所选材料和允许最大挠度的限制。在传统的砌体结构中通常采用受压过梁形式（比如采用水平拱）。此时拱结构的支座需要同时承受竖向和水平荷载。

　　钢筋混凝土过梁可以与楼板同时在现场进行浇筑，或者采用预制过梁的形式。预制过梁包括预制钢筋混凝土过梁、混凝土抹面钢筋加强预制砖墙过梁，U 型墙体预制过梁以及滚轴遮阳设备预制过梁等。

遮阳系统　　对于一些特殊的建筑朝向和外墙，有时需要设置遮阳系统（比如滚轴遮阳系统、窗帘等），而遮阳系统可能会安放到过梁内部或者附在过梁表面。此时，遮阳系统的安装会对过梁的承载作用造成一定影响，进而影响过梁对外墙的保护作用。

　　墙体开洞也可以做到与楼板齐平，此时不需要设置过梁。这种做法会显得室内更加宽敞，同时采光量增加。在必要的情况下，可以将洞口上部附件的楼板用钢筋进行加强。

窗栏　　开洞的底部可以采用窗栏的形式。实体墙和固定窗户构件必须具有足够的高度来预防人从开洞处跌落。根据窗户在外墙中位置的不同，需要设置内部窗台和卧式覆盖；对于外部，这种设置更为重要。

　　通常将暖气片安装于窗栏区域，可以降低来自玻璃附近以及其他构件的冷空气对流。通常来说，暖气片的安装不能对外墙的承载

> **提示：**
> 　　想了解更多关于静力体系、承载性能以及支座的更多信息，可以参考本套丛书中的阿尔弗莱德·梅斯特曼（Alfred Meistermann）编著、文捷翻译的《承重结构》一书（中国建筑工业出版社，北京，2011 年）（征订号：20278）。

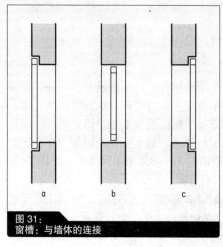

图 31：
窗槽：与墙体的连接

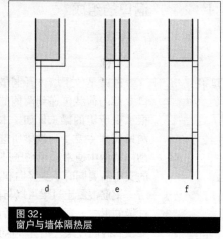

图 32：
窗户与墙体隐热层

力和热阻效率造成太大的降低。

一般来说,落地窗不设窗栏。此时落地窗底部的密封必须足够好,通常在底部增设一个门槛。当然,也可以通过设置格栅或者百叶窗,来实现通向外部或者阳台的无门槛通道。如果外部阳台或者门廊离地面的高度大于1m,则必须采用必要的措施以预防人们跌落。

窗槽

窗槽指的是窗框与建筑墙体之间的连接处,可以分为以下几种:〉见图 31

（a）设有内窗槽的窗户;〉见图 33

（b）无窗槽的窗户;

（c）设有外窗槽的窗户。

除此之外,窗户还能安放在以下不同的平面位置:〉见图 32

示例:

最大允许跌落高度

当窗户处于开启状态时,所需的窗栏高度与窗下实体墙以及固定玻璃构件的高度必须能够防止发生人跌落的情况。对于开启的落地窗,必须根据落地窗的开启方向在窗外或者窗内设置附加的保护措施,比如采用栏杆的形式。

窗下墙以及栏杆的设置高度取决于建筑的高度和建筑功能。在建筑12m高度以下的位置,窗栏的高度不得低于80-90cm（首层除外）,在12m高度以上的位置,窗栏的高度不得低于90-110cm。如果设有附加栏杆,窗下墙的高度则可以适当降低。

图33：
固定窗框周边设置的内窗槽

（d）与墙体内侧齐平；
（e）位于洞口中央；
（f）与墙体外侧齐平。

　　墙体的隔热层应该一直延伸到窗户所在的地方。通常在洞口的左右两侧和过梁三边设有窗槽。最常见的做法是设置居中放置的窗台和内窗槽——这种做法可以在墙体和窗框之间形成一个阻隔区，如果不设置窗槽，则需要在窗框宽度范围内形成节点（密封、隔热、固定等）做法。

　　通常在暴风雨较多的地区（比如北海海岸地区），通常采用外窗槽加窗台的做法。向外开启的窗户在风压力的作用下，由于窗槽和密封受压，窗户的整体性增强。

　　如今，选择窗槽形式的最关键因素是与窗户的安装方式。尤其对于多层建筑，当窗户发生损坏时，应该能够通过采用脚手架或者

外窗槽

提示：
　　阻隔区是一类具有特殊形状的结构构件，比如通过设置窗槽，在墙体结构内放置特殊构件，以防止水发生直接渗透。

重要：
　　保温层－窗槽
　　窗户应该安放于竖向保温层之内，以避免形成冷桥。窗槽可以通过设置阻隔区和保温层搭接的方式防止雨水渗透而发生聚集。

升降设备，从建筑外部进行安装和置换。如果窗户尺寸较大或者重量较重，则可以设置外窗槽来协助起重设备（起重机）工作。

如果窗户采用与外墙齐平的安装方式，则窗户的窗框、玻璃、洞口连接节点暴露程度较高，此时节点做法应该非常小心。从建筑物理的角度出发，由于露点温度在窗台和窗侧区域发生较大变化将导致冷桥的产生，那么即使在设置外窗槽（此时周边缝做成 U 型阻隔区）的情况下，要实现窗户与外墙平齐的做法也存在一定困难。在墙体内侧，则通过设置内窗槽并采用固定玻璃窗户的形式，实现窗户与内墙的平齐。

内窗槽

内窗槽一般适用于可在室内进行窗户安装以及维修的情况下。窗框从室内嵌入到窗槽中，然后向内安装直到整个窗槽宽度。过梁以及窗侧处的窗框宽度则取决于与窗槽的搭接宽度，最小可以减小至与开洞处框架剖面同宽。〉见图 33

由于内窗槽与窗框之间的搭接以及内窗槽的节点做法能够有效地形成阻隔区，所以内窗槽的安全性较高。窗户的安装能够做到与内墙平齐，此时与外墙平齐安装一样，必须采用必要的措施来预防冷桥的形成（比如，采用内保温层或者增加窗侧保温措施）。

无窗槽

如果在墙洞处无窗槽设置，那么将大大简化开洞处墙体的构造。但相应地对节点处的做法要求则大大提高——必须同时满足防风、防水汽以及无冷桥形成的要求。此时，与设有窗槽不同的是，此时只需要在窗户剖面宽度范围内满足相应要求。

只要窗户安装在保温层内部，或者在窗户周围侧向也采用了保温措施，那么就可以在墙厚范围内自由确定窗户的位置（与外侧齐平、与内侧齐平以及居中等）。不论从室内还是室外，均能够看到窗框的全貌。窗台的装修做法通常也不设凹槽，以确保节点处的功能要求和几何要求。

窗户构件的竖向荷载向下传递到下部承重墙上。窗户处的雨水应该采取措施向外排放，比如通过设置窗台。墙体截面——特别是墙体的附加保温层——尤其需要防水防潮。对于透气墙体，需要确保窗户底部空气的自由流通。〉见 144 页，图 57

在室内，在窗框底部必须采用必要的覆盖措施（针对设有窗下墙的情况）以及楼板设施（针对落地窗）。另外，在落地窗的底部还需要根据相关规范条文（比如，平面楼板规范条文）设置相应的防水密封做法。

P43　　**基本构造**

建筑竖向围护结构——立面——的开洞通常采用门或者窗户进行封闭。在外围墙体保护功能的基础之上，门、窗为墙体带来了开

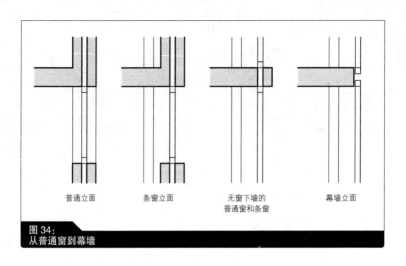

普通立面　　条窗立面　　无窗下墙的　　幕墙立面
　　　　　　　　　　　　普通窗和条窗

图34：
从普通窗到幕墙

敞和采光的功能。根据墙体构造的不同，窗框也在制作和断面做法上存在相应的不同标准。

普通窗户　　普通窗户的常见做法是在开洞处上方设有一道过梁，开洞两侧设置窗侧墙，开洞底部设置窗下墙。这种窗户做法可以形成连续窗或者长条窗，长条窗的侧面节点采用特定的窗户构件（可见框架）——设有顶部和底部的普通构件类型——形成。

玻璃幕墙　　对于玻璃幕墙立面或者窗墙立面，构件节点做法必须设置在顶部
立面　　或者底部。此时立面构件由不同的窗户单元形成，构件尺寸更大。竖向支柱和水平横杆将多单元窗户构件进行进一步划分。〉见图34和图35

开洞尺寸　　开洞尺寸将直接影响立面的建造方式和材料质量。墙体的构造与设计相互影响。高而窄的开洞适用于砌体墙中，较宽的开洞需要设置较高的过梁和洞口侧面的加强。通常认为，现浇钢筋混凝土墙体的承载力已足够（通过外部不可见的钢筋加强，墙体的承载力能

提示：
落地窗
防水做法应该做到覆盖层15cm以上高度的位置，以防止天气相关的水分从门槛以上渗透到室内。比如，如果发生雨水管堵塞或者雨水结冰等现象。如果条件良好，能够确保雨水随时通畅排出（比如，在窗洞外侧设置天沟），那么可以适当降低防水做法的高度，但最小高度不得低于覆盖层以上5cm。

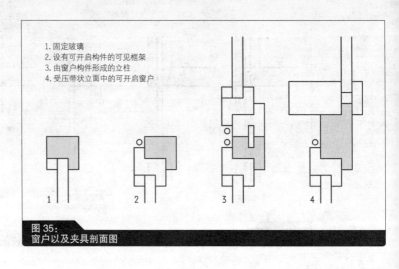

1. 固定玻璃
2. 设有可开启构件的可见框架
3. 由窗户构件形成的立柱
4. 受压带状立面中的可开启窗户

图 35：
窗户以及夹具剖面图

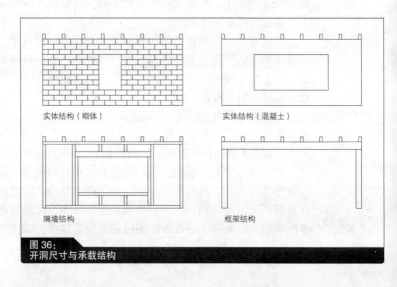

实体结构（砌体）

实体结构（混凝土）

隔墙结构

框架结构

图 36：
开洞尺寸与承载结构

够得到进一步地提高），这也意味着洞口的尺寸可以相应地增大。对于框架结构，开洞能够达到梁柱之间的空间（对于横墙承重结构，可以达到墙体与楼板之间的空间），而开洞的具体尺寸可以通过具体的需求分析确定。)见图36

如果开洞宽度范围内受到了阻隔或者固定玻璃的高度太高，则可以将较宽的窗户替换成窄条的条窗。这种情况下，可开启的窗扇

与墙体的连接

需要设有各自的窗框。所有的大型制造商都可以提供多种尺寸多种材质的窗户以及条窗,部分制造商已经实现了标准化(详见"窗户的组成"一章中"窗框"部分内容)。

连接窗户与墙体时,其侧面、上部与底部需要采用不同的方法。在窗下墙位置的连接区域所承受的荷载最大,而过梁处则可能因为需要设置滚轴卷帘或者遮阳系统使原本简单的几何做法变得复杂。)见图37 和表1

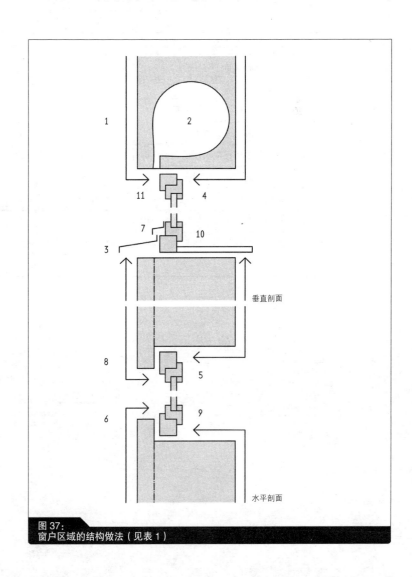

图37:
窗户区域的结构做法(见表1)

表1中详细介绍了窗户节点做法上的一些总体结构要求,比如荷载传递、保温层连续、足够的耐久性、不同结构构件之间的密封性、阻隔区、搭接要求等。

开洞设计概念要求（见图37） 表1

		两侧	上部	底部
1	开洞处、过梁处的荷载分布		×	
2	整体控制设备	×	×	(x)
3	外部条带边缘		×	×
4	耐久性、框架与墙体之间的密封	×	×	×
5	窗户的固定（无张力）	×	(×)	
6	将外部墙体覆盖层铺设到窗户构件上	×	(×)	(×)
7	外部预防大雨的密封做法	×		×
8	窗槽的设置	(×)	(×)	
9	将内部墙体覆盖层铺设到窗户构件上	(×)	(×)	
10	改变窗侧的材料（密封处、窗框处）	(×)	(×)	×
11	窗侧区域的保温层	(×)	(×)	(×)

×——始终需要

（×）——可能需要

窗户的组成

> 平开、旋转、
> 平开 – 旋转
> 窗户

开启方式

窗户具有多种不同的开启方式（详见"窗框"一节和图38至图40）。

平开窗可以绕竖轴向内或向外开启，而平开 – 旋转窗户在此基础之上还能够绕底部的水平轴向内开启。为了使用方便，平开、旋转、平开 – 旋转窗户通常向内开启。此类窗户的另一种变换形式是平开 – 折叠窗户，这种窗户可在室内开启向外折叠。〉见图41 这种窗户需要确定采用左侧还是右侧铰接的方式。

> 上悬窗或者
> 上悬外开窗

上悬外开窗向外开启，其转轴水平放置于窗户顶部，以防止窗户开启的时候雨水进入室内。〉见图42

对于可开启窗户，窗槽根据窗户开启方向的不同相应设置于窗框的外侧。〉见图43 顶部窗框与开启窗扇直接的节点做法应该考虑大雨的情况，通常采用在节点处设置挡水条确保雨水排出，防止雨水沿立面流动。

> **提示：**
>
> 在进行窗户开启方式的表达时，应该采用比例尺为1∶15的施工平面图。其中在外侧视图中转动应该采用连续实线（向外开启）或者虚线（向内开启）表示，推拉滑动应该采用箭头表示其开启的方向。在平面图中进行开洞尺寸标注时，需要标注洞口的高度与宽度，其中宽度尺寸标注在尺寸线上方，高度尺寸标注在尺寸线下方。
>
> 关于平面图表示的更多信息，可以参考本套丛书中的贝尔特·比勒费尔德（Bert Bielefeld）与伊莎贝拉·斯奇巴（Isabella Skiba）编著，吴寒亮、何玮珂翻译的《工程制图》一书（中国建筑工业出版社，北京，2011年）（征订号：20282）。

> **提示：**
>
> DIN标准EN 12519中有关窗户的规定：
> "左侧"：窗户沿铰链向左手侧开启，逆时针关闭。
> "右侧"：窗户沿铰链向右手侧开启，顺时针关闭。

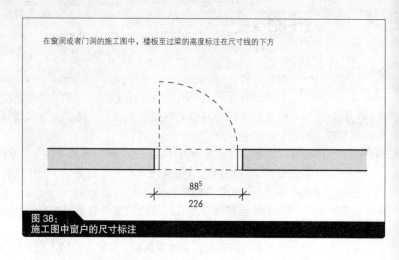

图38:
施工图中窗户的尺寸标注

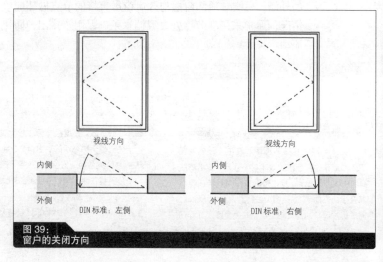

图39:
窗户的关闭方向

推拉窗与提升–推拉窗

推拉窗可以向侧向滑动,考虑到密封效果,通常选用内侧的窗扇进行开启。提升–推拉窗需要先进行竖向提升再进行侧向滑动,确保滑动起来更加容易。

竖向推拉窗通常采用两个窗扇、向上开启的形式,但对于大型窗户且不设窗下墙的情况时,也存在向下开启的情况。)见图43 有些情况下,活动的窗扇可以向下推拉至楼板以下的位置(比如,进入地下室楼层内),此时人们可以通过开启的窗扇出入于室外首层地面与

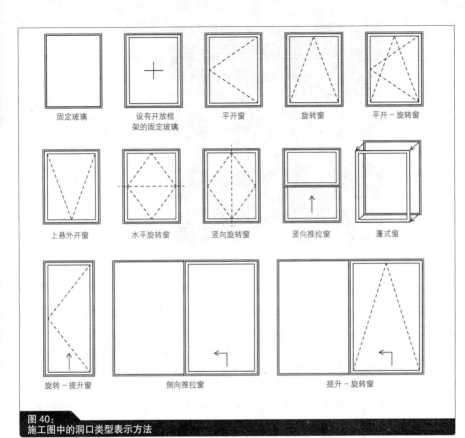

图40：
施工图中的洞口类型表示方法

图41：
折叠-转动窗户（金属）

图42：
上悬外开窗（铝合金）

图 43:
竖向推拉窗（木质）

	室内之间。
提升–推拉–旋转窗	提升–推拉–旋转窗还可以绕水平轴向内旋转，这些功能大大增加了相关设备的复杂性以及窗户所占用的空间。
水平旋转窗	水平旋转窗可绕其中间轴旋转，比如窗户的外侧面可以旋转的室内，以便于清洁和其他工作。由于旋转轴设置在窗户中间位置，那么窗户的剖面位置和窗槽位置需要相应发生偏移。同时，不论什么情况下窗户的最大出入面积只有窗户面积的一半，所以在将旋转窗作为逃生通道设置时需要特别注意。这类窗户在20世纪50年代和60年代的应用非常广泛，在如今已经很少见，其主要原因是旋转轴附近位置的密封难以保证。
竖向旋转窗	竖向旋转窗与水平旋转窗的工作原理相同，存在的问题也一样。竖向旋转窗沿着中央的竖向轴旋转开启。
板条窗	板条窗由一系列小型条板或者竖向、水平的可转动条板组成。如果每个条板都具有自己的框架，那么窗框与玻璃的面积比值将很高。每一块条板的尺寸在 200×100mm 至 2000×400mm 之间，而窗户的尺寸则在 300×150mm 至 2000×3000mm 之间。
箱式窗	箱式窗需要通过一系列的装置将窗户向外推开至墙体立面之外，所以这种窗户的通风效果非常好：冷空气可以从窗户下方流动，而热空气则可以从窗户上部排出到室外。当处于开启状态时，窗框任何一边都没有和窗洞进行可靠连接，所以箱式窗的外伸设备需能承受所有的荷载。

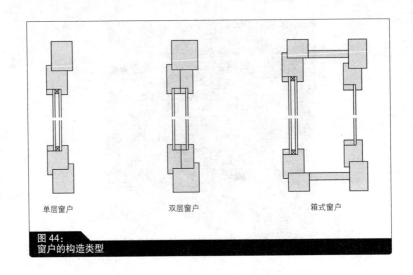

单层窗户　　双层窗户　　箱式窗户

图 44:
窗户的构造类型

P52
单层窗户

双层窗户

箱式窗

构造类型 〉见图 44

由于隔热窗玻璃的保温效果较好，单层窗户是目前的标准窗户类型。设有隔热窗玻璃的窗户的传热系数大约在 3.0W/（m²K），而设有隔热窗玻璃和窗框的窗户的传热系数大约在 1.3W/（m²K）。

在发明隔热窗玻璃之前，单层窗户的传热系数 U_w 大约为 4.8W/(m²K)，此时采用双层窗户则可以大大改善其保温隔热和隔音效果。双层窗户通常采用两个窗框的组合形式，比如同时进行开启。通常窗扇之间的距离为 40-70mm，这种做法的隔热效果比采用隔热窗玻璃的单层窗户稍有提高 [其传热系数 U_w 大约为 0.3W/（m²K）]。

双层窗户截面可打开进行清洗。双层窗户通常采用平开型、倾斜型以及平开内倾型窗户。

在一些老式建筑中，通常还可以看见由两层窗户简单组装成一体

提示：

整体传热系数

在评价保温隔热效果时，窗户结构需要当成一个整体进行评价。其评价指标为传热系数 U_w 值。U_w 值（之前也称为 k 值）表示的是整个结构的节能能力。U_w 值越低，通过窗户的热量损失越小。U_w 值可以综合考虑玻璃的传热系数（U_g）、窗框的传热系数（U_f）、玻璃周边的热传递系数以及各构件所占的面积比值综合考虑。

图45：
箱式窗（钢窗）

的窗户形式。这种窗户比现代双层窗户的厚度大很多，而且其组合做法通常不太牢靠。如今，内层窗户一般采用了隔热玻璃，两层玻璃之间的中空部分增加了隔热和隔音的效果。如果两层窗户采用前后排列并采用连续衬板（通常采用木质）连接成整体的形式，那么我们称这种窗户为箱式窗。箱式窗通常采用平开的开启方式。见图45

箱式窗通常采用19世纪常见的方式，用窗闩对窗户进行划分。当窗框以及窗闩所占比例越高，以及窗扇之间间距越大，则损失的光线越多。现代附加玻璃立面做法的前身就是箱式窗。由于箱式窗的制作工艺复杂、成本高，目前主要集中应用在对隔音要求非常高或者翻修指定使用的情况下。

无框窗

对于现代窗户，其节能效果最差的地方通常是窗框而不是玻璃。为了提高节能效果，一种可能的方法是减少窗框在窗户中所占的比例。无框窗的最大特征是在窗户的一侧设置不可见窗框或者其他窗户固定设备（比如点支或者夹支设备）。玻璃采用了阶梯形刃口玻璃：外侧玻璃伸出至内层玻璃边缘之外，然后与支撑框架黏结，而外层玻璃后面形成了一条连续的黑色黏结带。

大多数的开洞方式都可以采用无框窗的形式，但无框窗的黏结节点与夹支节点相比在施工和维修方面要求更高。

P55

固定窗框

窗框

固定窗框伸入到墙体结构内部中，而窗户可开启部分的框架与

窗框相互连接，或者在固定窗框内安装固定玻璃。〉见图46
　　（1a）木质窗框，固定窗框的顶部
　　（1b）固定窗框的底部（图46中未显示）
　　（1c）固定窗框的侧部
　　（2）立柱（窗间柱），将固定窗框进行竖向划分（图46中未显示）

图46：
窗户的组成部分

（3）横档（窗闩），将固定窗框进行水平划分

在安装固定窗框时，需要确保窗框尺寸与洞口尺寸相互匹配，误差在可允许范围内，以防止在窗框构件以及墙体中造成过大的拉力或者挤压。窗框与建筑的连接应该做到稍有弹性或者可自由滑动（详见"结构构件组装"一节"密封"部分内容）。

窗扇框架 窗扇框架是窗扇的一部分，其与固定窗框连接并可以开启。见图46

（4）窗扇框架

（4a）木质窗扇框架，窗扇框架的顶部

（4b）木质窗扇框架，窗扇框架的底部（挡水板）

（4c）木质窗扇框架，窗扇框架的侧部

（5）横档，将窗扇框架进行水平划分

（6）玻璃

玻璃落地窗 具有两处可开启区域的窗户可以采用立柱（窗间柱）进行划分，比如采用固定窗框中的窗间柱。或者，可以采用升高的装饰做法，比如盖缝条作为窗户可开启区域的组成部分，从而覆盖该区域的接缝以及密封处。这种做法通常被称为玻璃落地窗。

材料——系统窗户

木窗

木窗的最大优势在于其保温隔热性能良好（导热系数对比：云杉——$\lambda=0.11W/mK$；铝合金——$\lambda=209W/mK$），操作简便以及材料的可持续性。木窗不需要复杂的制作过程、制作成本较低而且不需要对窗户框架进行复杂的保温隔热处理。

在木窗的制作中主要采用针叶林树木（比如松树、云杉、黄杉和冷杉等），落叶林树木（比如橡树、刺槐）使用的较少，同时对于热带落叶林树木（比如梅兰蒂木、红木、非洲柚木）的使用则越来越少。窗框的表面要经受紫外线照射、干燥、雨水等日常天气的作用，

重要：

可允许误差范围对窗户的名义尺寸进行了规定，包括尺寸大小、形状以及结构构件与建筑的位置等。"误差"还包括了设计时与理论尺寸的预期偏差，这种预期偏差考虑了包括窗户材料以及制造商等因素在内可能引起的潜在偏差。

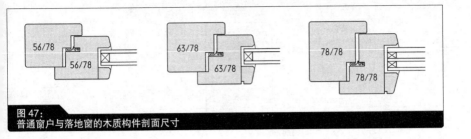

图 47：
普通窗户与落地窗的木质构件剖面尺寸

木材的含水量通常在 25%-50% 之间，而这种湿度增大了腐败菌、霉菌以及昆虫对木材的损害。所以说，适用于窗框的木材必须具有足够的强度以及较低的吸湿性和天然抗病性。〉见图 49

对于原木窗户，需要采用特殊的表面处理方式进行保护。刷底漆能够有效地防止引起木材褪色的霉菌，而浸泡处理则可以防止湿度引起的腐烂。在此，需要区分刷清漆与刷涂料覆盖层的区别。需要注意的是：颜色越深，其吸热升温效果越明显。采用斜边或者圆角边缘能够确保涂料与木材之间的黏结能够更加持久。

在制作木窗时，更重要的是对受力结构构件的保护。必须防止积水的浸泡，同时需要确保雨水从木材表面流走，比如，在木窗中应该避免水平表面的设置。

断面

木材横截面具有标准尺寸和等级（标准断面）供选用。〉见图 47 简要标注以毫米为单位标注了断面的宽度和高度。通过采用实木窗框（比如，采用三层夹合木），其 U_f 值可以降低至 $1W/(m^2K)$。

构件尺寸

窗框的断面尺寸直接影响窗扇的最大尺寸，而二者均与窗户的开启方式有关。比如，当窗框断面尺寸为 56/78 毫米组合时，考虑到自重和转动时对铰轴产生的荷载，采用转动窗户的窗扇高度不得大于 80 厘米，而落地窗（转门）的宽度不得大于 1 米。〉见图 48

P57

金属窗

金属的导热性很好。钢铁的传热系数大约是木材的 250 倍，而铝的传热系数大约是钢铁的 4 倍。所以，在窗户构造中，金属框架必须采用有效的隔热保温措施。

市场上有大量的铝合金窗以及钢窗供选用，而断面尺寸差异较大。与木窗不同的是，金属窗没有固定的标准尺寸。

铝合金窗

铝合金窗的断面隔热层通常采用挤压成型的半成品材料。内外

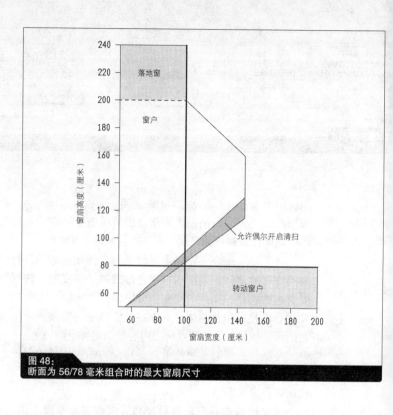

图 48：
断面为 56/78 毫米组合时的最大窗扇尺寸

框架断面之间采用塑料板或者硬质泡沫塑料进行隔离。尽管铝合金窗成本投入较高（由于其原材料生产过程中的耗能量较大），但与同尺寸的木窗或者塑料窗相比，铝合金窗仍然具有一定的经济性。铝合金窗的使用年限更长，而且其保养和维修更加方便。另外，铝合金窗的加工尺寸偏差较小，比如窗框能够非常精确地下料安装。同时铝合金窗的工作性能较好，重量也较轻。〉见图 49

在铝合金窗与建筑相连接的地方应该能比木窗或者钢窗承受更大的温度引起的长度变化（在温差为 60K 的情况下，单位长度的铝合金长度变化将达到 1.5 毫米）。在设计时，在固定窗框与建筑之间以及内部大尺寸窗户和立面构件之间所采用的节点应该能够承受足够的温度变化。

一般来说，铝合金窗的表面需要进行涂层处理，未处理的铝合金容易发生不均匀氧化和生锈。在此，需要区分机械表面处理（磨光、刷光、抛光）与电化学表面处理（电镀，能够在表面形成一层均匀

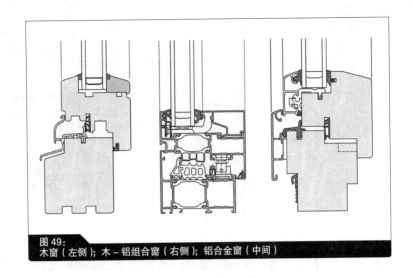

图49：
木窗（左侧）；木–铝组合窗（右侧）；铝合金窗（中间）

的氧化膜）。另外，铝合金表面还能够进行烤搪瓷处理或者粉末涂层处理。粉末涂层处理指的是将需要涂层的材料放入到喷粉机中，然后放置于炉温为180℃左右的烤炉中进行喷粉处理。

双层窗户（木–铝合金窗）

木–铝合金双层窗户很好地发挥了两种材料各自的优势。木质窗框的导热系数较小，故放置于内侧；铝合金层的承载性能以及耐候性较好，故放置于外侧用作装饰和防止受天气影响的作用。

在木材与铝合金连接处必须采用滑动连接，以防止两种材料之间的温度变形差异，同时必须防止在二者搭接位置发生冷凝现象。

木–铝合金双层窗户存在多种不同的结构类型。通常，可以在标准尺寸的木窗外附加铝合金覆盖层。另外，还可以在木窗外侧设置铝合金窗槽并进行整体密封。这种做法可以将窗框的导热系数 U_f 降低至 1.3–1.5W/（m²K），而对于设有中间隔热层的多层木窗窗框（比如，在之间设置聚氨酯泡沫塑料），其导热系数 U_f 可达到 0.5–0.8W/（m²K）。

钢窗

钢结构窗框可以由热轧T型、L型或者特殊形式的型钢直接加工制成。目前，最常见的是由高强钢板冷轧成形的中空截面形式。钢结构窗框的抗弯、抗扭刚度以及承载能力均比铝合金窗户高。窗框由一系列保温的截面组成。)见图50

钢窗的最大弱点是腐蚀问题，可以通过涂保护漆、电镀、采用不锈钢截面等方式来进行预防。钢窗也可以进行烤搪瓷或者喷粉处理。

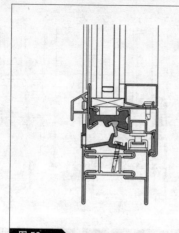 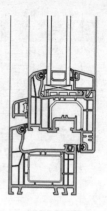

图50:
钢窗（左侧）；塑料窗（右侧）

P60
PVC 截面

塑料窗（PVC：聚氯乙烯，GRP：玻璃增强塑料，玻璃钢）

硬质 PVC（聚氯乙烯）可以通过挤压形成单腔或者多腔截面体系。其中，多腔体系的保温隔热能力更强。在进行窗框制作时，单根的挤压成型截面通常采用斜向焊接的形式。PVC 截面的导热系数很低，但其强度和承载能力也很低，所以通常在塑料截面的空腔体内设置金属构件用来提高窗框的稳定性，此时由于金属构件位于空腔体内部，所以不需要进行隔热处理。〉见图 50

在最外侧的空腔中需要通过排气孔安装冷凝水排水装置和水汽扩散补偿装置。大多数的 PVC 窗户为白色，也可以进行染色或者涂层处理，但不能刷漆。

由于 PVC 的温度膨胀系数较大，在太阳光照射下会发生较大的纵向膨胀，对于深色塑料现象更加明显。另外，太阳光照射还会造成 PVC 颜色的变化。

GRP 截面

为了避免窗框部分造成的过多的热量损失，可以改善窗户的 U_w 值，也可以采用 GRP（玻璃增强塑料，玻璃钢）截面作为窗框材料。GRP 的导热系数较低，同时还具有较高的强度和刚度，所以不需要另外的加强措施。此外，GRP 截面可以和铝合金截面组合使用。

P60

玻璃系统

玻璃系统主要包括玻璃、安装槽口以及与窗框之间的密封条。

设计窗户时，在考虑窗框的类型、材料的同时，还需要重点考虑玻璃系统的性能需求。主要包括以下几方面：玻璃的保温性能、玻璃的尺寸、对玻璃的特殊功能要求（保温隔热性能、防火性能等）、玻璃的安放以及密封过程等。〉见图51

在进行窗户的节能评估时，除了考虑窗户的温度导热系数 Uw（用 W/m^2K 来表示）之外，另一个需要重点考虑的是窗户玻璃处的能量吸收。能量吸收指的是热量传递的效率或者获取的太阳能（g 值）。g 值可以直接通过太阳能传递以及附加热传递（其他长波热辐射或者对流）计算得到。玻璃系统的 g 值可能用 0-1 或者 0%-100% 的数值进行表示，数值越高，通过玻璃进入室内的能量越多。

具有多种不同质量、不同性能的玻璃种类可供选用。

单层玻璃

浮法玻璃

20 世纪 50 年代后期，玻璃制造商 Pilkington 发明了一种新的平板玻璃制作方法——浮法。液态玻璃从熔炉中进入到 1100℃的大罐中，然后较轻的玻璃溶液从熔罐中溢出流入到两边设有平行挡板的带板中。根据设备的不同，平板玻璃的最大长度通常在 6-7m，厚度为 1.5-19mm。

影响玻璃板宽度的因素较多，包括运输允许的最大尺寸。如果玻璃板采用低架拖车进行运输，那么过桥允许的 4m 最大高度、交通

重要：

浮法玻璃的物理性质，TSG（全钢化安全玻璃）和 HSG（热增强玻璃）：

容重：	$2500kg/m^3$
弹性模量：	$70000-75000N/mm^2$

温度性质：

导热系数：	$0.8-1.0W/(mK)$
U_w 值：	$<5.8W/(m^2K)$

声学与视觉性质（厚度范围为 3-19mm）：

声音衰减指数评估值：	22-38dB
透光率（Lt）：	0.72-0.88
辐射透过性能：	0.48-0.83

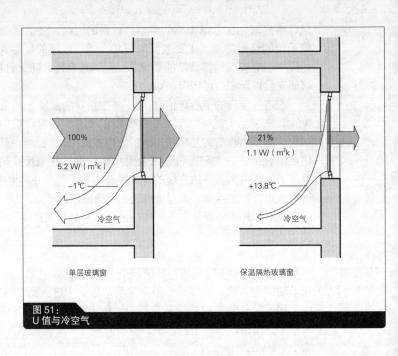

图51:
U 值与冷空气

窗户 U_w 值示例（2004 年） 表2

	U_w 值（W/m²K）
单层玻璃窗	4.8
夹层中空隔热玻璃窗（厚度单位为 mm）(4/12 空气层 /4)	3.0
夹层中空隔热玻璃（4/12 空气层 /4）	2.7
三层中空隔热玻璃（4/10 空气层 /4/10 空气层 /4）	2.2
三层中空隔热玻璃（4/8 充气层 /4/8 充气层 /4）	1.7
双层镀层中空隔热玻璃（4/20 充气层 /4）	1.3
三层镀层中空隔热玻璃（4/10 充气层 /4/10 充气层 /4）	0.9
双层窗户 / 箱式窗	2.3
特殊隔热玻璃	0.4
防盗玻璃	1.6

图52:
玻璃板，透明的、半透明的

灯的高度以及运输车的转弯半径都是影响玻璃宽度的关键因素。减去运输车的高度，也就是说玻璃板的宽度大约在3m左右。

根据制造商的不同，常见的浮法玻璃宽度通常为3.2m，而最大长度约为7m。在生产的过程中，还可以对浮法玻璃进行染色处理。浮法玻璃所具有的天然绿色调，可通过特殊原材料的调整，生产出"白色玻璃"以及"超白玻璃"。通过摩擦、蚀刻以及喷砂等手段，可以将天然透明浮法玻璃加工成半透明玻璃。浮法玻璃破碎时将产生大块锋利的碎片。浮法玻璃可以通过进一步的加工，形成更多形式的产品。）见图52

压铸玻璃　　在生产压铸玻璃的过程中，液态玻璃熔液通过若干对滚轴挤压形成一面光滑、一面具有纹理，或者两面均有纹理的玻璃。玻璃表面的纹理能够散射光线，形成半透明的效果。压铸玻璃的破碎形式与浮法玻璃相同。在压铸的过程中，能够生产出特殊形状的玻璃（比如：U形玻璃）。

嵌丝玻璃　　嵌丝玻璃在玻璃中增设了一层丝网，丝网通常采用不锈钢材料，并在交叉点进行焊接处理。在生产的过程中，将丝网压入到玻璃熔液中。当玻璃发生破碎的时候，丝网能够将玻璃碎块拉在一起，防止伤人。通常（需要参考各地相关规定），嵌丝玻璃设置在头顶以上部分。水可能沿着嵌丝玻璃边缘向内渗透，从而引起金属丝网的锈蚀，或者在内部结冰引起玻璃的剥落。因此，嵌丝玻璃的边缘需要进行密封处理或者设置框架进行保护。

P63
钢化安全
玻璃

热处理玻璃

首先将玻璃加热到转化温度,然后立即进行冷却(比如:在玻璃上方吹冷空气),将在玻璃表面形成压应力,属于预加压应力。

这种做法将改变玻璃的组织结构,并且提高玻璃的抗弯承载力和热冲击承载力。当玻璃遭到严重损坏的时候,玻璃将破碎成细小碎块,并形成紧密网格结构,并且小碎块的破碎角均为钝角。在经过热处理之后,钢化安全玻璃不能再进行机械处理。

热增强玻璃

热增强玻璃(又称半钢化玻璃),其表面预应力能够确保玻璃损坏时,仅发生边缘至边缘的穿透性裂纹。热增强玻璃(HSG)并不属于安全玻璃。其生产过程与全钢化安全玻璃(TSG)相似,但在吹冷空气冷却的过程更为缓慢一些。所以,热增强玻璃与全钢化玻璃相比,更容易发生破碎。与普通浮法玻璃相比,热增强玻璃的抗弯曲破坏承载能力和抗热冲击性能更好。

P64
夹层玻璃

多层玻璃

夹层玻璃由至少两层玻璃板以及玻璃板之间的夹层(薄膜或者树脂)组成。但夹层玻璃并不能满足任何安全需求。

夹层安全玻璃(LSG)

能够满足安全需求的夹层玻璃具有至少两层玻璃板。所以,夹层安全玻璃通常采用浮法玻璃+全钢化玻璃或者双层热增强玻璃等做法。两层玻璃板采用树脂或者片状PVC薄膜(聚乙烯醇缩丁醛树脂)作为中间层连接成整体。夹层薄片具有较好的弹性和较高的抗裂承载力,并且可以进行染色或者研光处理。夹层材料填塞在两层玻璃之间,然后在两层玻璃的压力之下进行加热融化。在持久潮湿的环境下,PVC材料的黏结性能将削弱。因此,在作为隔热玻璃应用时,必须确保玻璃安装在可自由通风的清洁玻璃槽中。在发生破损时,玻璃碎片不会进入到夹层中,所以即使在破损的情况下,夹层安全玻璃仍然能够满足安全要求。

> 重要:
> 对于全钢化安全玻璃(TSG),在完成温度处理之后不能进行机械加工,故在钢化之前应该确定可能的钻孔或者切割尺寸并完成加工。

P65
绝缘玻璃

功能玻璃

绝缘玻璃通常是由至少两块相互分开、周边连接的玻璃组成。在玻璃之间的空隙中保留干燥空气或者填充其他气体，用来改善玻璃的隔热或者隔音功能。玻璃的周边密封对保证玻璃的气密性至关重要，通常在玻璃采用垫圈以及一道主密封（比如丁基橡胶）和一道次密封（聚硫橡胶）。将垫圈置于丁基橡胶中，然后与洁净的玻璃板黏结，再覆盖另外一层玻璃板，最后采用聚硫橡胶进行二次密封。

垫圈位于内侧的一边设有凹槽，凹槽内填设干燥剂。典型的绝缘玻璃结构应是内侧为6mm厚浮法玻璃，中间为12mm中空层，外侧为6mm厚浮法玻璃。见图53和表3

P65
隔热玻璃

特殊玻璃

隔热玻璃的隔热性能的提高主要是通过在两层玻璃之间充入特殊气体或者是采用3层而不是通常的2层玻璃板 [U_g值为 $0.5W/(m^2K)$]。根据充气情况的不同，玻璃之间的间距通常会增大10-16mm。

玻璃板之间的填充气体具有较差的导热性能（氪气的U_g值最低，为$0.5W/(m^2K)$，采用$2 \times 12mm$的分隔）。同时，隔热玻璃的玻璃板采用中性颜色的稀有金属进行涂层覆盖以减小导热系数，涂层的位置为内侧玻璃板的外表面。

吸热玻璃

吸热玻璃的主要功能是防止玻璃后方的空间发生温度过高，从而不需要在玻璃背后设置其他可能需要的遮光设备。吸热玻璃能够允许大部分的可见光入射，同时阻止大量产生热量的光线入射。

在工作方式上，需要对以下两种方式进行区分，但有时这两种方式也同时发挥作用：

— 吸热玻璃是指在玻璃上涂加金属氧化物涂层，涂层能够吸收

提示：

不同制造商对于隔热玻璃的标准尺寸

根据制造商的不同，隔热玻璃的最大尺寸在$420 \times 260cm$与$720 \times 320cm$之间，其总重量可达到3.5吨，隔热玻璃的最大尺寸由可生产的最大面积决定。6-10mm厚的全钢化玻璃可生产的最大面积可达到$10.9m^2$。隔热玻璃的最小尺寸如下：浮法玻璃组合——$24 \times 24cm$；全钢化玻璃组合——$20 \times 30cm$。根据所需性质的不同（隔热、吸热、隔音等），涂层工艺也存在差异，也将导致玻璃尺寸的差异。

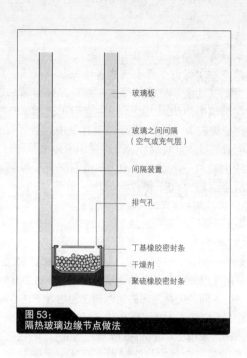

图 53:
隔热玻璃边缘节点做法

建筑节能规范对玻璃性能的要求（2006 年 3 月） 表3

	U_g 值 W/m²K	g 值 %	Lt 值 %
单层玻璃（Pilkington Optifloat clear 玻璃）	5.8	85	90
设 10–16mm 中空层的双层隔热玻璃	3.0	77	80
4/16/4 充氩气双层隔热涂层玻璃（Pilkington Optitherm S3 玻璃）	1.1	60	80
6/16/6 蒸汽涂层太阳能玻璃（Infrastop Brilliant 平板玻璃）	1.1	33	49
4/12/4/12/4 充氪气三层隔热涂层玻璃（Pilkington Optitherm S3 玻璃）	0.7	50	72
玻璃砖	3.2	60	75
Pilkington Profilit 单剖面玻璃（翼缘宽度 60mm）	5.7	79	86

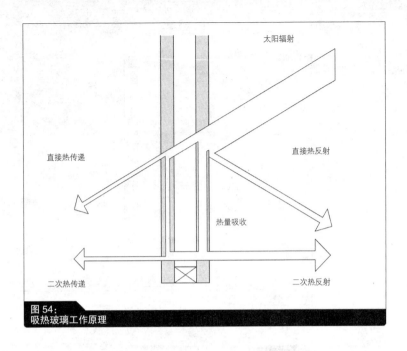

图 54：
吸热玻璃工作原理

入射太阳能，然后转化为热能。转化的热量中大部分向外散射，只有很少部分热量在滞后一段时间后向内传递。〉见图 54
— 反射玻璃通过金属氧化物涂层将大量入射太阳能（紫外线以及红外线）进行反射，同时允许大量的可见光入射。通常，涂层设置在外层玻璃板的内表面上。〉见图 55

可开关和变化玻璃

如今，随着技术的发展，可开关和变化的玻璃得到了重视。这种玻璃具有特殊的结构形式、可开关的结构层以及特殊的激活方式（电致变色、气致变色、光致变色以及热致变色等）。通过改变玻璃处的电流、气体控制、温度以及太阳光，可以控制进入玻璃的太阳光比例。通过这种方式，可以根据具体的天气情况、季节以及时间对入射光线进行控制和调整。

隔音玻璃

玻璃的隔音效果可以通过大致的声衰减指数（Rw）进行表示。通过以下方式可以改善隔热玻璃的隔音效果：

— 增加玻璃板的重量；
— 内外采用不同的玻璃板厚度（形成非对称结构）；
— 采用复合玻璃板；

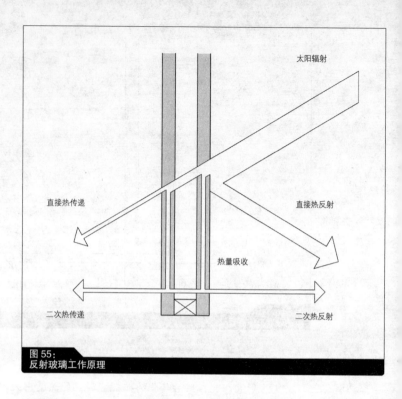

图 55：
反射玻璃工作原理

> 🔍 💡
> 防火玻璃

— 增加玻璃板间距并进行充气处理。

在讨论防火玻璃的时候，必须同时考虑玻璃框以及玻璃框与墙体的连接方式。防火玻璃可以分为 G 型和 F 型两种。

G 型玻璃可以在一定时间内防止火苗和可燃气体的侵入，但不能阻止热量的辐射。比如，G 型玻璃可以采用全钢化玻璃或者与淬

> 🔍
> **示例：**
>
> 声衰减系数通过一个单一值定义结构构件或者房屋之间的隔音效果，该值与结构单元的自振频率相关。我们需要区分一下两种空气声衰减系数：
>
> R'w：采用分贝（dB）为单位表示包含声音传递的结构构件的隔音效果；
>
> Rw：采用分贝（dB）为单位表示不包含声音传递的结构构件的隔音效果；
>
> 典型结构构件的空气声衰减系数：
>
> — 墙体，天花板　　R'w
>
> — 门，窗　　　　　Rw

重要：
窗户的隔音值
在德国，根据窗户的隔音效果进行了如下分类：

隔音效果分类	窗户做法	大致声衰减系数
1/2 类隔音效果	设置隔热玻璃的单层窗户（4/12 空气层 /4）	25–34dB
3 类隔音效果	设置隔热玻璃的单层窗户（8/12 空气层 /4）	35–39dB
4 类隔音效果	设置隔热玻璃和树脂填充的单层窗户（9/16 树脂填充 /6）	40–44dB
5 类隔音效果	设置隔热玻璃和树脂填充的单层窗户（13/16 树脂填充 /6）	45–49dB
6 类隔音效果	双层窗户：隔热玻璃（9/16 空气层 /4）以及单层玻璃（6mm）	45–49dB
	箱式窗户：隔热玻璃（6/16 空气层 /4）以及单层玻璃（6mm）	>50dB

火钠钙玻璃形成的复合玻璃来制作。

　　F 型玻璃在防止火苗和可燃气体侵入的同时，也能够防止热量辐射。若在防火区域的背后存在防火逃生通道，则需要采用 F 型玻璃。F 型玻璃的玻璃板之间填充了凝胶，如果在火灾下发生破坏，则凝胶会转化为坚硬的实体泡沫。

　　固定玻璃可以达到 F90 的耐火极限等级。通常，在选用 F 型玻璃时，需要根据相关的建筑法规对 F 型玻璃的数量、类型以及与墙体的连接部位进行检验。

防盗玻璃

　　防盗玻璃可以根据其阻隔等级进行分类（0–6），包括用于银行和邮局柜台的特殊类型。防盗玻璃可以与防盗警报系统组合使用。安全玻璃可以起到防盗，有时甚至是防弹的功能。防盗玻璃的等级

示例：
DIN 标准 4102 对建筑材料及构件的防火性能规定——阻燃玻璃

结构构件	耐火极限（分钟）
墙、楼板、梁、柱	F30–F180
楼梯、窗户/玻璃系统	F30, F60, F90, F120
	G30, G60, G90, G120
防火构件（门，大门，盖板）	T30–T180

是根据机械斧破开边长为 400mm 的开洞所需次数来确定的。

紧固系统

通常采用镶玻璃条将玻璃板安装到窗框中。玻璃板应该安装在窗内侧以防止被打碎，同时玻璃板还应该能够被拆除，以确保在玻璃发生破碎的情况下能够进行更换。镶玻璃条能够将玻璃板和密封条连接成为一个整体，并且能够承受水平荷载（比如，风荷载），然后将荷载传递给窗框等承载构件。

镶玻璃条应该确保在玻璃板平面内产生压应力，以防止玻璃发生破碎。根据窗户体系的不同，玻璃条可以采用螺丝拧紧或者夹紧的方式与窗框连接。

窗棂

固定玻璃和可开启窗扇均可以采用窗棂进行分格，这种做法与玻璃的发展历史有关。在最初的时候，只能生产出较小的玻璃板，通过设置窗棂可以将较小的玻璃板布置到较大的窗户面积中。如今，玻璃生产主要采用浮法工艺，不再需要设置窗棂。但是，从保留传统的角度出发，在涉及相关的建筑法规或者当地的建筑特征时，仍然保留了窗棂的设置。

玻璃的安置——垫块

在安装玻璃时，需要在玻璃与窗框之间设置垫块来分散玻璃的重量。我们需要区分承载玻璃垫块和非承载玻璃垫块：承载玻璃垫块设置在窗框中，发挥支承玻璃的作用；非承载玻璃垫块的设置主要是为了确保玻璃板边缘与固定窗框之间的间隔。对于固定玻璃，玻璃荷载直接传递到固定窗框中，而窗框与墙体之间进行锚固。对于可开启窗户，玻璃荷载需要通过可开启窗框以及窗框的支承结构（比如铰链、滚轴等）进行传递。对于正常工作的窗户，需要确保可开启窗扇在开合过程中不发生阻碍、扭曲或者变形等情况。玻璃板与窗框之间不能在任何位置发生接触，同时要确保玻璃板之间以及玻璃与安装槽之间的空间保持均匀一致。

玻璃板与安装槽之间的蒸汽压必须与室外空气的蒸汽压保持平衡，以避免发生蒸汽冷凝的情况。排气孔的设置需要确保不直接承受风荷载的作用，有时候可能需要设置盖板来保护排气孔。另外，玻璃板与安装槽之间的蒸汽压不能与室内空气的蒸汽压保持平衡，否则将会在间隔处发生冷凝水聚集的现象。

控制设备

控制设备指的是窗户体系中所有控制窗户开合、安全、固定以

图 56:
窗户的控制设备

及使用等功能的机械部分。⟩见图 56 窗户以及落地窗的控制设备将固定窗框与可开启窗扇二者进行结合，窗户的开启设备必须具有防盗、防止儿童操作以及防风和防雨水渗透的功能。

控制设备通常是成组提供的。控制设备的安装能够做成完全隐蔽（不可见）、半隐蔽或者完全外露（比如，具有装饰效果）的形式。打个比方，对于平开 – 旋转窗户，窗扇与固定窗框之间的连接通常采用铰链（主要为钻孔铰链）和以下附加设备：

——撑杆：控制设备的组成部分，在窗户开启或者关闭的过程中进行安置或者移除。通过推拉撑杆能够控制窗扇的开启或者关闭。

——转动轴：窗户旋转的支撑点，同时承受窗扇的重量。

——剪刀铰：安装在固定窗扇的上部，与窗扇设备相连接。与窗户转动轴联合工作，剪刀铰能够决定窗户的旋转轴，同时控制窗户平开与悬开之间的转换。

——锁片：锁片安装在固定窗框一侧与窗锁接触。通过锁片，即使在恶劣的天气下，也能够确保窗扇的密闭状态。

不同的开启方式对应各自不同的控制设备。

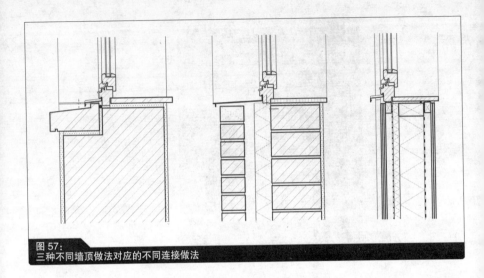

图57:
三种不同墙顶做法对应的不同连接做法

P73　**结构构件的整体安装**

窗户底部
窗下墙的外侧水平窗沿通常采用窗台板的做法。为了确保其耐久性和避免遭到损坏,必须注意以下几点:
— 窗户剖面必须延伸至窗台板内部,确保二者之间存在搭接并设置披水条,以确保在暴风雨情况下的防水性能;
— 窗台板必须伸出外墙外侧,确保20mm以上搭接宽度并设置滴水挑檐,以防止雨水污染墙体;
— 窗台板必须具有一定的向外倾斜角度(至少为5°或者8%),以便于雨水流出;
— 窗台板应该设置上翻的边缘(比如,设置金属条或者搭接做法),以防止墙体结构受潮。将其与墙体覆盖层之间设置横向搭接,也是保护窗台板和窗侧板之间连接节点的有效做法;
— 可能需要注意窗台板的温度膨胀变化(最大允许长度)。

外窗沿做法　外窗沿可以采用镀锌板、铜板、玻璃、天然石或者人造石等材料,或者采用竖立迭砌的抗冻天然石/人造石、缸砖、劈离砖,以及预制铝合金型材等。

内窗沿做法　窗下墙内部或者散热器的覆盖通常采用砂浆砌筑天然石或者人

图 58:
落地窗的平整底部做法

造石的做法，或者采用木材以及木材衍生产品（在需要的地方设置龙骨）进行覆盖。〉见图 57

落地窗　　　与设有窗下墙的窗户不同，落地窗自上而下均采用窗框支撑。在必要的地方，应该设置防潮和防水沥青、阳台板覆盖台阶，以及与室内地板设施有效连接。对于可开启的落地窗窗扇，其底部的水平构件截面通常比普通窗户大，其主要目的是防止雨水溅起弄脏玻璃。如果与实心门结合使用，可以通过一些措施，比如在底部设置窗槽的做法，确保二者底部构件在同一水平面上。〉见图 58

P74　　　**密封做法**

窗洞口的密封做法可以防止雨水从窗户处渗入到建筑内部，同时能够减少不可控的空气交换带走的热量损失。在此，需要区分以下三个不同地方的密封做法：固定窗框与墙体连接节点处的密封做法（压缩密封）；固定窗框与可开启窗扇之间的密封做法（企口密封）；窗框与玻璃之间的密封做法（接缝密封）。

压缩密封　　　压缩密封能够平衡建筑物与固定窗框之间的压力，同时对二者之间进行密封，抵抗室外的风、雨、噪声和室内的水汽扩散。窗户与建筑物之间在压缩密封区域进行机械连接，连接构件承受和传递竖向以及水平荷载。在建筑物洞口以及窗框形状发生变化时，压缩密封不能发生损坏。〉见图 59 与图 60

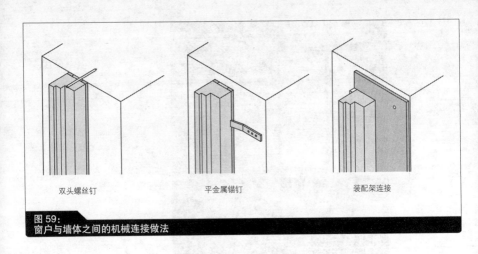

图 59：
窗户与墙体之间的机械连接做法

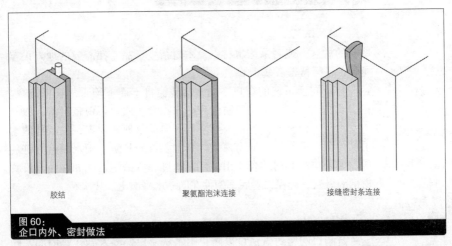

图 60：
企口内外、密封做法

企口密封　　固定窗框与可开启窗扇之间的企口密封应该具有一定的隔音隔热功能，同时能够抵挡湿气和气流。企口密封由密封条（比如，合成橡胶或者氯丁橡胶等）构成。密封条与窗扇中的槽口黏结或者压入到槽口中，同时还在固定窗框的对应位置设置密封条。密封条可以采用单层或者双层甚至三层密封的做法，同时应该确保密封条便于更换。

接缝密封　　窗框与玻璃板之间的接缝密封采用连续线条的形式，在玻璃板

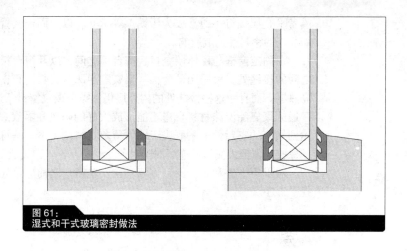

图61:
湿式和干式玻璃密封做法

两侧均有设置。接缝密封可以采用玻璃密封条（由硅树脂、丙烯酸酯、聚硫橡胶以及聚氨酯材料制成的密封剂）等湿式密封的做法，也可以采用预制密封条（比如合成橡胶等）的干式做法。>见图61

预制密封条采用挤压加工方式，所以能够生产出各种不同的复杂截面形式。与湿式密封不同，干式密封的安装能够在玻璃板垫块安装的同时完成。

接缝

窗户所承受的竖向荷载通常传递到窗户下方设置的垫块或者水平放置的楔子上，水平荷载则由窗侧板区域的窗框销钉、衬垫螺钉

> **提示:**
> 窗框与墙体之间的连接:
> 窗框与窗侧板之间的生产安装误差可以通过弹性压缩密封条吸收，压缩密封条通过销钉连接具有类似滑动连接的功能，或者采用U型框架连接而具有调节功能。通常立面材料或者窗户生产厂家将结合洞口尺寸提供准确加工的产品，所以在窗户设计和制作的时候可以允许具有一定的尺寸误差。

或者平金属锚钉（通常成为压铁）等承受。以上固定装置不能对压缩节点的密封性造成影响。

在固定窗框与砌体墙企口、窗台板之间，以及窗台板与窗下墙之间的接缝处，可采用矿物棉、聚氨酯泡沫等材料进行密封，形成隔热层。而且，这些接缝处的内外层仍然需要设置密封条进行密封。普通的聚氨酯泡沫材料自身不能形成太好的防风和隔汽效果。为了达到最好的密封性——取决于墙体覆盖材料的类型——可以采用丁基橡胶在窗框外侧形成连续的密封条。

对于不平整和多孔的砌体墙，在安置窗户之前应该先进行抹灰找平工序。

示例：

预压密封胶条（压缩条）：在聚氨酯基底上预压浸渍泡沫密封条，压缩量大约为截面高度的15%，在安装到位之后压缩条缓慢膨胀。膨胀条与接缝边缘应该紧密连接，以适应不同材料在温差作用下的变形差异。

重要：

窗户安装中的相关工艺

U 型阻隔区

防止暴雨和湿气的渗透。设置于窗框剖面中：通过在窗框或者固定窗框中设置单企口或者双企口的方式。设置于墙体中：通过不同结构构件之间的搭接（窗侧板/窗槽与窗框搭接，窗框与可开启窗扇窗框之间搭接，密封条与窗台板以及窗台板与墙体之间搭接）在墙体内侧或者外层形成企口以防止雨水的聚集。

建筑运动允许差异

窗框与墙体之间：采用可调节的弹性安装方式以及连续的胶状接缝剂。窗框与固定窗框之间：采用企口缝的做法。

隔热

将窗户安装于墙体隔热层平面内，采用隔热玻璃，在压缩密封处填充矿物棉或者聚氨酯泡沫等。

热分离

在窗框与玻璃之间进行热分离。对于全木质窗框以及附加窗扇的金属窗框或连接处采用塑料条；塑料窗框截面中设置成型金属截面；为隔热玻璃设置垫片以及在玻璃板之间设置中空层或者充气。

密封

在窗框与玻璃之间：采用密封条以及密封材料的二次密封；在窗框与墙体之间：通过设置企口，连续的金属箔片以及接缝密封条等（见图62和图63）。

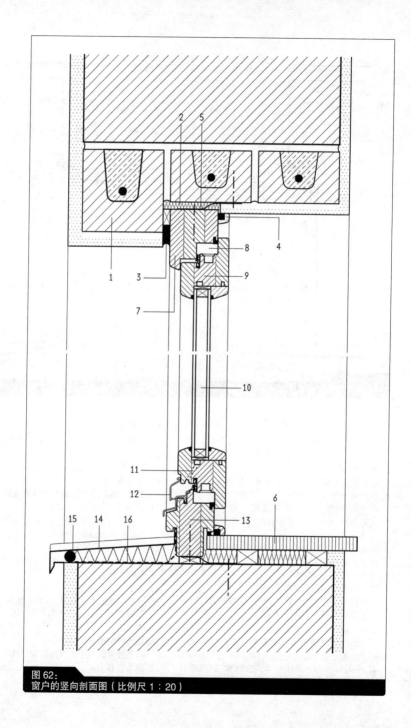

图 62:
窗户的竖向剖面图（比例尺 1∶20）

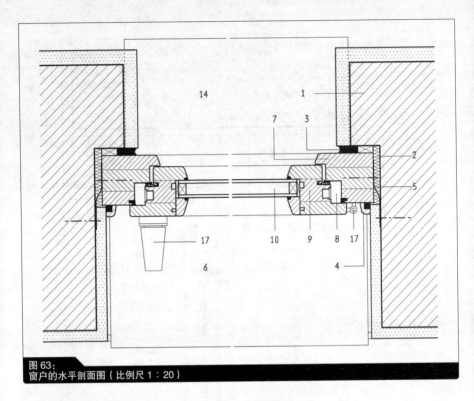

图 63:
窗户的水平剖面图（比例尺 1：20）

1. 墙体连接——内企口
   窗过梁：此处为 L 形墙体
   窗侧板：砌体外墙形成三边企口（顶侧边）
2. 企口隔热
   填塞矿物棉或者塑性泡沫（聚氨酯泡沫）
3. 企口密封
   内部隔汽密封：在环形底层（聚氨酯泡沫）上涂胶状密封剂（比如硅胶）；外部：采用防风密封条（压缩条）
4. 盖条
   保护和密封压缩密封节点
5. 盖条
   保护和密封压缩密封节点
6. 窗户衬板 / 窗台板，滑移安装、螺丝拧紧
7. 固定窗框。截面：IV 78/78
8. 窗槽空隙
   用于安放相关设备和关闭元件
9. 可开启窗扇窗框
   截面：IV 78/78，采用整体式窗槽和 窗框密封做法。窗框中在玻璃板四 周开槽并包围玻璃板。
10. 隔热玻璃
    8/16mm，玻璃板间设 4mm 中空层
11. 蒸汽压补偿
    通过钻孔确保窗槽空隙的通风和排水（冷凝水）
12. 挡水条
    湿气和冷凝水可以通过挡水条底部的开洞排出
13. 底部固定窗框
    通常设有"滴水挑檐"。与外窗台板进行搭接，防止接缝处遭到雨水渗入。
14. 挡水条 / 外部密封
    具有至少 5°的向外坡度，挡水条通常采用方形截面铝合金材料。
15. 底部窗槽密封
    在玻璃密封条上设置胶状密封剂（比如硅胶），保护外部人造石窗台与组装 / 固定窗框之间接缝处防风条。
16. 防风胶带形成的防风节点
    防风胶带设置于无窗槽的窗户四周。
17. 相关设备

结语

立面开洞是一个热点的建筑课题。窗户是影响节能设计和采用太阳能辅助加热、自然采光通风等环境友好型建筑的一个关键因素。

对于建筑师而言,老旧建筑的翻新是其工作内容的一个重要部分。此时,洞口的设计对于节能和设计效果尤其重要。现存大量的建筑希望能够在第二次世界大战后的几十年间得以继续使用,但是否能够实现这个愿望,则常常仅取决于建筑的窗户形式。从数量上来说,历史建筑或者建筑群中的开洞较少,但从质量上来说却具有更高的建造要求。

但是,并不是所有的设计都仅仅针对正确的尺寸或者合理的比例设计。窗户作为一种多功能构件,不仅需具有建筑物所有的传统功能,还会增加新的功能。打个比方,布鲁诺·陶特(Bruno Taut)受到许多区域建筑的启发,在他位于柏林的"Onkel-Toms-Hütte"建筑(1926-1931年)中,给(厨房)窗户的概念增加了新的变化。每个窗格均设置两个可开启窗扇,其中一个是竖向的矩形窗扇,另一个则是宽而矮的形式,二者采用相互叠放的形式,从而形成了镜像的效果。这种设计并不仅是从形式上吸引人,而且能够通过错列的小型窗扇确保连续的通风。

如今,窗户在机械、装饰和风格上的发展均与窗户设计的改变密切相关,从而给窗户印上了时代的特征。从建筑领域的角度出发,洞口的设计和建造需要考虑结构、设计等方面的因素。不论是对于老建筑还是新建筑,洞口设计都具有突出的重要性。

附录

参考文献

Francis D.K. Ching: *Building Construction illustrated*, 3rd edition, John Wiley & Sons, 2004

Andrea Deplazes: *Constructing Architecture*, Birkhäuser Publishers, Basel 2005

Martin Evans: *Housing, climate and comfort*, Architectural Press, London 1980

Gerhard Hausladen, Petra Liedl, Michael de Saldanha, Christina Sager: *Climate Design*, Birkhäuser Publishers, Basel 2005

Thomas Herzog, Roland Krippner, Werner Lang: *Facade Construction Manual*, Birkhäuser Publishers, Basel 2004

Ernst Neufert, Peter Neufert: *Architects' Data*, 3rd edition, Blackwell Science, UK USA Australia 2004

Christian Schittich, Gerald Staib, Dieter Balkow, Matthias Schuler, Werner Sobek: *Glass Construction Manual*, Birkhäuser Publishers, Basel 2007

Andrew Watts: Modern Constuction: *Facades*, Springer, Vienna, New York 2004

图片鸣谢

Figure page 8 (Th. Herzog)	Peter Bonfig
Figures 1, 2, 3, 4 (Ingenhoven Overdiek), 5 (F. O. Gehry), 6 (K. Melnikov), 7, 8, 13, 14, 18 (Ch. Mäckler), 20, 22, 23 (A. Reichel), 25 (P. C. von Seidlein), 30 (Kada Wittfeld), 41 (Kollhoff Timmermann), 42, 43, 45 (W. Gropius), 56 (Brinkman & van der Vlugt), page 78 (Wandel, Hoefer Lorch + Hirsch)	Roland Krippner
Figure page 12 (Lorenz & Musso), 24 (Lorenz & Musso), 33 (Lorenz & Musso), 52 (Lorenz & Musso)	Lorenz & Musso Architekten
Figure page 36 (Herzog & de Meuron), Figures 9, 10, 12, 31, 32, 33, 34, 35, 36, 37, 38, 39, 40, 44, 47, 48, 51, 53, 54, 55, 57, 61, 62, 63, Tables 1, 2, 3	Department of Building and Building Material Studies, TU Munich
Figures 11, 16, 17, 21, 26, 27, 28, 29	Herzog et al., Facade Construction Manual, 2004
Figures 15 (Sieveke+Weber), 19 (Sieveke+Weber)	Sieveke+Weber Architekten
Figure 46	Sonja Weber
Figure 49 left	Meister Fenster IV 72/75, Unilux AG, D-54528 Salmtal
Figure 49 right	Aluvogt, Bug-Alutechnik GmbH, D-88267 Vogt
Figure 49 centre	Hueck/Hartmann Aluminium Fensterserie Lamda 77XL, Eduard Hueck GmbH & Co. KG, D-58511 Lüdenscheid
Figure 50 left	Janisol primo, Jansen AG, CH-9463 Oberriet
Figure 50 right	System Quadro, Rekord § Fenster + Türen, D-25578 Dägeling
Figure 58	Fa. R&G Metallbau AG, CH-8548 Ellikon a. d. Thur
Figures 59, 60	Krüger, Konstruktiver Wärmeschutz, 2000
Work on drawings	Peter Sommersgutter

作者简介

罗兰·克里普纳（Roland Krippner），工学博士，建筑师，德国慕尼黑工业大学工业设计系学术助理。

弗洛里安·穆索（Florian Musso），教授，工学博士，德国慕尼黑工业大学建筑工程与建筑材料研究系全职教授。

学术及编辑助理：索尼娅·韦伯（Sonja Weber）、硕士工程师，托马斯·伦岑（Thomas Lenzen）硕士工程师，德国慕尼黑工业大学建筑工程与建筑材料研究系。